Cover Your A$$ets

Asset Management at Your Place and at Your Pace

John L. Ross, Jr.

INDUSTRIAL PRESS, INC.

Industrial Press, Inc.
32 Haviland Street, Suite 3
South Norwalk, Connecticut 06854
Phone: 203-956-5593
Toll-Free in USA: 888-528-7852
Fax: 203-354-9391
Email: info@industrialpress.com

Author: John L. Ross, Jr.
Title: Cover Your A$$ets: Asset Management at Your Place and at Your Pace
Library of Congress Control Number: 2019938039

Printed in the United States of America.

ISBN (print): 978-0-8311-3641-3
ISBN (ePDF): 978-0-8311-9521-2
ISBN (ePUB): 978-0-8311-9522-9
ISBN (eMOBI): 978-0-8311-9523-6

Editorial Director/Publisher: Judy Bass
Copy Editor: Janice Gold
Compositor: Patricia Wallenburg, TypeWriting
Proofreader/Indexer: Claire Splan

industrialpress.com
ebooks.industrialpress.com

10 9 8 7 6 5 4 3 2 1

For Mom and Dad

Contents

Foreword

I met John Ross a few years ago at a training session for planning and scheduling, and have continued to stay in touch with him. John is an out-of-the-box thinker who uses humor to get his point across. He has a unique, fun, and energizing writing style that draws the reader in. When I read his work, I find it difficult to put the book down—a rare thing when dealing with any technical subject. This engaging style helps John to clearly communicate winning techniques to deliver a zero loss reliability environment.

When reading his books, it feels like John is sitting across from you, in your office, having a spirited discussion. You will get the feeling of having your own consultant on call, adjusting the advice depending on your own personal situations. More importantly, when you are done with the book, you will have a strong strategy to follow and implement. I highly recommend you read this work.

Ron Rieger
Global Mars Manufacturing Excellence Leader
July, 2019

Acknowledgments

I want to start by acknowledging and thanking those of you who bought my first book, *The Reliability Excellence Workbook: From Ideas to Action*, and making it such a huge success. I have received much praise for that body of work and many clients that I work with have purchased copies. That means a lot to me personally, so I thank you. I am humbled that so many would honor me by purchasing and reading my initial work. Many of these folks I still work with and I continue to enjoy the discussions the book has generated.

I also want to extend my appreciation to my many colleagues in the consulting business who have purchased the book and have positively remarked on my work. I have enjoyed hearing other points of view that might be different from my own. I've been very consistent when I say that "if you have a way to do it, and I have a way to do it, we now have two ways to do it." Please keep all the comments and well wishes coming.

In the context of that first book, I would like to give special thanks and recognition to the educational division of the Marshall Institute. The Director of Training Services, Tom Furnival, and the Instructional Designer, Angie Faucette, immediately saw value in my work and commissioned that a Reliability Strategy Workshop be developed and made part of Marshall's 2019 (and beyond) training calendar. This is an exceptional hands-on and transformative course where you, the learner, actually leave the workshop with a working reliability strategy to implement at your facility. I am proud of my work and more proud to be associated with Marshall Institute.

This acknowledgment section would not be complete if I did not thank the numerous skilled and support personnel that I've had the pleasure to work with over my three (plus) decades in maintenance and reliability. To say that I've worked with some superstars would be an understatement. I mentioned to a friend recently that one way to identify successful leaders is to see if their people get promoted. I can proudly proclaim that many people that have worked for me in the past have gone on to achieve higher levels of success. I would like to think that I played a small part in their success, but to be honest, these folks were exceptional before I even met them. I was just blessed to be associated with them. I've learned a lot from the people around me. I learned by listening. I've been known to say, "I learn more when I'm not talking." To them I can only say thank you. Much of what I will convey in this book is based on our interactions and our conversations. I am better for listening to my many knowledgeable friends and colleagues.

Judy Bass and the outstanding group at Industrial Press should be highlighted for believing in me as an author, and me as a professional in this space. Judy and her associates, Janice Gold and Patricia Wallenburg, are experts and more importantly, friends. Thanks as well to my personal editor, Mary Jo Richards, for her constant reminders of proper English and punctuation.

As a parent, I am always mindful that the young people my kids 'hang with' are the first level of folks to influence them and apply peer pressure. Many of you reading this acknowledgment section can attest to the fact that you are concerned if your child is running with the wrong crowd. My kids are in their 20s and they ran and are now running with a good crowd. My kids are simply great. I'm proud of Bailey and Austin, and I appreciate their support in my work and in my life.

I'm mentioning this as a lead-in to thanking my own first level of influence, my own close friends and family. My personal 'inner group' is positive, uplifting, caring, and genuine. I am thankful to count a group of childhood friends as part of my personal extended family. I also count their parents as close friends of mine as well. I think their lifelong friendship to me has given me a great sense of confidence, enough to write a book or two.

My family is simply awesome. I wrote about them in my first book, but I will expand on that to just mention how much I respect and love my sister and brother and their families. We didn't grow up on the rich side of town, but our parents always made it known that we could be anything we wanted to be. This was a great environment to grow up in. As a result, both of my siblings are greatly supportive of me and my ambitions. I return that sentiment every chance I get. Thank you, Regina and Mike.

There are many people who are responsible for much of the text and content of the book you are about to read. Believe it or not, some of them I've never met. These are the authors, content creators, thinkers, and doers in an asset management environment who have gone before me to lay the tracks for me and others to follow. It is my intention to respectfully cite and reference their previous work and either agree, add to, or promote another thought altogether. I think we are missing the value of debate in our industry. We have replaced the honest and robust dialogue of respectful dissention with arguments, passive aggressiveness, and mandates for absolute compliance.

The people I will quote and the works I will cite have kept the debate alive. Their work inspires me. I may not know them, but I thank them and you should too.

It was my great fortune to be born into a loving family and the luck of the draw to have such professional associations for my own professional growth. Coupling that with a gentle push over the finish line has contributed to what I hope you will find to be a ready resource to help you *Cover Your A$$ets.*

I Just Don't Think You're Smart Enough—An Introduction

"We keep saying we have no other course,
what we should say is we're not bright enough
to see any other course."

—Dan Carlin quoting David Lilienthal,
Atomic Energy Commissioner,
The Destroyer of Worlds podcast

I'm troubled. It's more like I'm concerned, really. Honestly, I already feel better for telling you this up front, although I do realize that this is an odd way to start a book. You need to know that I am writing this under duress. No, I'm not in any physical pain or jeopardy. No one is forcing me to write this book. It's just that I wonder if we are capable of doing what we are being asked to do. After all, we haven't done it yet and that troubles me. Could it be that we just aren't smart enough?

Starting a Book, This Book

I told a friend recently that I had started this book in my head eleven times. "Eleven," she said. "That seems like an odd number." "It is," I responded. I wanted to tell her that eleven was also a *prime* number, but I didn't want to seem pretentious (it's an engineering thing).

The perplexing issue with the process of writing this book was getting my opening salvo just right. I wanted to create the perfect thesis, or argument that would bring the reader in. I wanted that one *hook* that would result in half of the readers proclaiming, "Oh, hell yeah!" and leave the other half asking, "What the hell?"

The thesis statement is key to piquing the interest of the reader. I think the message in this text is important, and I know this book will help you with asset management at your place and at your pace. The goal is, through typed words, to connect the message to the intended recipient. Let's face it, a good book that is not read has the same effect as no book at all. I wanted to write a compelling argument. It was the way I was taught and the way I went on to teach others.

Years after attaining my Ph.D., I began teaching as an adjunct faculty member for a few local colleges and universities. My first graduate class was a graduate level research course. Completing this research course was a requirement for students before writing and defending their Master's Thesis. I told my class that one of the first things they needed to give consideration to was developing their thesis statement. "Have a bold and compelling statement," I told them. "Have a thesis statement that triggers an immediate response from the reader. Nobody wants to read a *vanilla* paper. Average papers don't invoke passion and don't get read. Get your audience to disagree with you or boldly agree."

To get my class in the mindset of writing an awesome thesis statement, I held weekly debate sessions. Just short twenty-minute discussions. Short but impactful.

Each week I would introduce a current event topic and tell the class to be ready to debate it the next week. I divided the class into pro and con groups. When we met again, I said something along these lines, "Guys, I decided to switch you up. If you were the con side of the debate I'm going to ask you to argue 'for' the issue, and vice versa." This went on every single week. It didn't take my class long to realize that in order to make an effective argument, for or against, you need to be able to argue both sides of the debate. Know your competitor's *stuff* better than they know it was my point.

How did the chapter heading of this introductory chapter make you feel when you read it?

"I Just Don't Think You're Smart Enough"

If I did my homework, that statement made you a little curious as to what was coming next. Let me tell you where that line actually came from.

It's Not Rocket Science

One of my first consulting assignments was at a sugar factory in the Deep South. I was there with the president of Marshall Institute, Greg Folts. Prior to this assignment, I had been wrestling with a nagging thought. Remember, I was brand new to the consulting gig. My quandary was, "How does a maintenance manager feel when their boss calls in a maintenance consultant?" I had no experience with this. I had only ever worked with one consultant, and I was the one who called him in.

Through the mastery of his years of knowledge and experience, Greg completely and sincerely laid out a Total Productive Maintenance (TPM) strategy for the potential client. This approach would absolutely deliver what the client wanted and even provided some things Greg knew they would have to have, but they had yet to discover that they would need. On the way out of the plant manager's office, the maintenance manager lightly grabbed my arm and leaned into me and whispered, "Please help me."

Here's what I discovered right then and there. The maintenance manager and the plant engineer almost always know *what to do*. Their biggest obstacle is convincing everyone else that it is the *right thing to do*. We are going to get that alignment right during the course and content of the book that you now hold. The twenty-first century is not a time for maintenance versus production. This is a time for asset management, and that means everyone!

Years later and now a more confident consultant, I was alone and pitching a plan to a plant leadership group. The plant manager was clearly not a person who valued teamwork and felt that all things equipment reliability related were solely within the jurisdiction and responsibility of the maintenance department. At the end of my presentation, he said, "I just don't think we can do that." Admittedly a little frustrated, I said, "You know, I agree. I just don't think you're smart enough."

That woke some people up.

The plant manager explained that he didn't mean that they weren't smart enough to do what I was suggesting, it was more of an issue of 'his' willingness to engage production in what was traditionally a maintenance responsibility. He didn't think production would do what I was asking them to do. Production was not motivated, in other words, and thus, not capable. This was the first group to ever hear my famous "Unless it requires us overcoming the laws of physics—inertia, the speed of light, or gravity—it is all within our capabilities" speech.

After all, it's not rocket science.

Give that point some serious thought. Unless the solution to our problem requires us to overcome the laws of physics, everything we need to be successful is within our control. It's often no more, really, than time and money.

If we say we *can't* do something, we are really saying that we aren't smart enough to figure out *how* to do it.

A New Standard Is Born

The epiphany you should be experiencing at this time is to have the metaphorical scales lifted from your eyes and see that almost anything we will be asked to or required to do is within our capabilities. We just need to come together in our organizations and facilities to understand what is being asked, and determine how we are going to respond. The good news? We always seem to have something to practice on because we always seem to be evolving.

Our newest challenge is to understand and implement ISO 55000, Asset Management. ISO is the moniker for the International Organization for Standardization, a global fraternity of national standards bodies. ISO 55000 isn't new, although with this title it has only been around since the beginning of 2014. Remember that global fraternity I just mentioned a few sentences ago? The birth of this standard, ISO 55000, came from the publication of PAS 55 by the British Standards Institution. This new standard has some history of application and success and almost nothing in the documentation can be argued against, although we may try. The paradox is that it all makes such common sense. As we know, common sense isn't always that common.

But haven't we seen this movie before?

I purposefully started this chapter by mentioning that I was troubled and concerned. I went further to reveal that I was writing this book under duress. Let me explain.

Why Isn't Anyone Listening?

I've been in maintenance for a long time, and I have the hairline to prove it. In fact, I would put my bona fides up against anyone's to prove that I am confident in what I say and do. I spent eleven years as an aircraft maintenance officer in the United States Air Force, sixteen as a plant maintenance manager or plant engineer, and seven as a reliability consultant. I believe I've consulted or worked in every industry you could name and on almost every continent. No, I take that back. I've never worked in the whaling industry in Antarctica. But I've been around.

Throughout my thirty-four-year career in maintenance and reliability, there has been a common thread. This thread ran through the military, private industry, and almost all the plants and facilities I've consulted in. What is this common denominator? Constantly convincing operations and organizational leadership that we need help with equipment maintenance and reliability. We need access to assets to perform proactive and corrective maintenance. We need money and resources to do our jobs. We need production to stop running equipment into the ground. Why doesn't anyone else care about the state and the upkeep of the company equipment? Why isn't anyone listening?

Imagine the surprise of all professional reliability and maintenance experts when they realize that all we needed to do was to publish an international standard to get people hopping to our rescue! Read that last sentence with a dramatic eye roll and as sarcastically as possible.

I asked earlier, haven't we seen this movie before? You don't have to go as far back as W. Edwards Deming to begin to see the concept of asset management taking shape. Seiichi Nakajima alluded to our combined interest in asset performance and reliability in his introductory book on Total Productive Maintenance as recently as 1972. Of course, there have been numerous offshoots of this work, some as recent as 2018 (a point that is relevant depending on when you read *this* book).

The point? With all the knowledge of all the maintenance managers and maintenance supervisors, and all the collective wisdom of plant engineers and reliability engineers, why is there only interest now on a grand scale for asset management? Because ISO 55000 is not a maintenance book, and it's not written by maintenance people. It is for those 'other guys.' Please, please, please keep this in mind. ISO 55000 is not a maintenance book and it is not written for just maintenance people. It is time for others to step up and get involved in asset management.

My fear, and the honest reason I am concerned? Your boss, or your boss's boss, is going to assemble a committee, read through ISO 55000, 55001, and 55002 and decide that maintenance needs to up their game. The 'brass' might think that in order for your company to be in compliance with the ISO 55000 standards it will take more work from the maintenance department and it will quickly become another maintenance program.

But wait, now we know this isn't the way to execute asset management, and we should be smart enough not to go down this road again. After all, we know how that movie will end.

ISO—My Personal Journey

In early 1995 I started my civilian career at a manufacturing plant in rural Illinois. I was hired to be the plant engineer, maintenance manager, and maintenance supervisor. This was a metal working facility; they made copper-bottom pots and pans. I'm sure you know the company.

Shortly after I began, our parent company, Corning, committed our company to becoming ISO 9000 certified within a year. There was a lot of office buzz about what ISO was, and what 9000 was all about. I had absolutely no idea what ISO was, and I couldn't research it easily because the Internet didn't really exist in 1995. Not in my little plant it didn't.

We were told that the ISO 9000 certification was a means to 'vet' us to our customers. Through this quality standard, our customers could essentially do away with any incoming inspections for quality. And likewise, we would no longer have to inspect incoming raw production principal supplies from our vendors who were also ISO 9000 certified. I don't think it ever really worked out that way to be honest, but we were successful for our part.

For maintenance, our ISO certification meant that we had to spin up a calibration lab for the tools that the tool and die makers used to make our die sets. Think about that for a minute. We were a global company, a wholly owned subsidiary of Corning (a huge company). We made the die sets in our machine shop that made the pots and pan products we manufactured. All this work was executed daily and we didn't even have calibrated tools for our die makers. How on earth did we even function back then?

I was responsible for creating an ISO 9000 certifiable tool calibration lab. Again, I was fresh out of the military and I had no idea what ISO meant. We did learn that the certification audit was going to simply be a review of our processes and a confirmation of whether or not we were compliant with our own processes. The mantra was "Say what you do, and do what you say." Ok, that was simple enough.

My plant was a union plant, so I had to post for a calibration lab technician and the job went to one of my machinists, a lady named Paula. I wasn't surprised; Paula's employee number was 3. The other two had died, so she was the odds-on favorite.

Paula did an exceptional job pulling this all together and the details she had to master as a tool maker rolled right into what we needed to keep the calibration paperwork straight. We got certified on our very first attempt. A vetting company came in for the audit and they were amazed at our processes and amazed that we actually did what we said we do. The audit cost $25,000, in 1996. It would have been a tough sell to have to repeat that audit.

I want to share two fun stories regarding our attempt to get ISO 9000 certified. I mention these stories for two reasons: they are interesting and fun, and they show that work can be interesting and fun. If you and your company are going to make a run towards ISO 55000 certification, don't forget to have fun and learn something along the way.

Paula and I discovered that we had to have 'standards' for our machine shop, and that those standards had to be tested on another set of standards, and those standards had to be traceable all the way back to the NIST (National Institute of Standards and Technology). When I say 'standards' I don't mean processes, I mean that we had to have steel blocks of exact measure to calibrate our micrometers and calipers against. Our steel blocks had to be confirmed and certified against the absolute exact standards of the United States.

Paula and I set out to find a calibration lab to partner with. We located one in Chicago and made arrangements to visit and possibly set up a contract for services. For some reason Paula asked if she could drive, and I was in no hurry to drive up to and in Chicago so I had no problem with her request.

On the way out of town, Paula mentioned that she needed some gas, so we pulled into the small gas station on Main Street. I got out to pump the gas, and noted that Paula had pulled *her* car up to the gas pump on the wrong side. The gas tank door was on the other side of *her* car. I told Paula this and she said, “Sorry, let me turn around.” Paula went on to execute the most beautiful (I am not making this number up) seventeen point turnaround you have ever seen. She pulled up to the exact same pump in exactly the same orientation. I had not moved one inch. “There,” she said. I shook my head.

In Chicago, we found an eager little calibration company that wanted our business. During a tour of their facility, my interest was captivated by a bell jar sitting on a pedestal in the exact center of their shop. Inside the container was a perfectly smooth, brush metal item roughly the size and shape of an old Zippo cigarette lighter. I was transfixed by this metallic object. I asked the owner what this metal item was and he told me that it was three inches. I asked, “Three inches of what?” “It’s exactly three inches,” was his response.

I had never seen such a thing in my life. I was looking at an object that was precise to 10 millionths of an inch. I’m still fascinated by that experience to this day.

Here is my hope for you, the reader. We may or may not know each other. I would sincerely love to meet and get to know everyone that is interested in advancing the profession of maintenance and asset reliability. We are a small fraternity so we need to work together, but my immediate hope for each of you is that you have great success in this journey towards asset management. Also, that you travel down *your* own path and at *your* own pace to become wildly successful. However, do it in the spirit of enjoyment, and have fun. Don’t forget to learn something along the way and share it with others.

Helpful Hints to Get the Most from This Book

In 2018, I completed a bucket-list level ambition and published my first book. My first offering exceeded my greatest expectations and I was very pleased with the work that I had created. The feedback from my first book was 100% positive.

It was during the creation of that book, *The Reliability Excellence Workbook: From Ideas to Action* that I hit on a method, a genre of book writing. In essence I had stumbled upon a workbook format that delivered exactly what I was hoping to achieve. At an SMRP (Society for Manufacturing and Reliability Professionals) annual conference in Kansas City, MO in 2017, I met the lady who would eventually become my publisher. When Judy Bass asked me to tell her about my book idea, this is what I told her: "I want to write a book that is as if you and I are sitting around a dinner table, drinking coffee, and we are just talking about stuff and I'm sketching out some thoughts on a napkin." That is literally the first conversation Judy and I had.

That first book delivered on my idea, completely. I wanted a book that allowed me to introduce an idea, ask what *you* thought about it and we would both generate some anecdotal stories to support the concept. I'd share some world-class principles with you and together we would start to piece together a comprehensive maintenance strategy for you at your location.

I'm adopting that same theme for this book. I have broken down the concept of asset management into the working elements that I believe are relevant. I am a constant learner, so I will interject points of view from other authors, making this a bit of scholarly work. I'm going to ask you what you think, and how things are back at your place. This will be a workbook of sorts because I want to give you space to jot down your thoughts. Throughout this text, we will be working side by side to put together the strategy for how asset management could and should be implemented at your location.

Mechanically, you will be asked to fill in the blank spaces. I'll highlight those spaces in light gray so you will have some indication as to where to record your thoughts.

I ask that you let others participate and record their ideas as well. I think everyone in your organization could benefit from this level of discussion and engagement. There is a lot to share and discuss.

The Good Stuff Inside

This is a very shareable book in the sense that we are going to touch on the major elements of asset management. It is interesting, and absolutely essential, to first determine what an asset is. This should surprise most of the people reading this book. If you truly gave it some serious thought, you would conclude that although we might label something as an asset, we certainly don't treat it like one.

Since this is a book on asset management, and we are focusing on industrial assets (manufacturing, service, and facilities), we will eventually be narrowing our discussions down to capital assets, essentially the equipment we utilize to provide the product or service. I call capital assets the 'stuff that makes the stuff.' That is an easy way to differentiate capital assets from human assets (the people who make the stuff) and financial assets (the money to make the stuff), and so on.

The aforementioned ISO 55000 standards will be reviewed in the text of this book. Much of the three books (55000, 55001, and 55002) are open for interpretation, similar to my story about my introduction to ISO 9000. Remember, "Say what you do and do what you say."

Once we've begun to put some framework around our interpretation and intent towards capital assets, we will create roles and responsibilities around the assets and begin to nip at an actual asset management approach.

Many of you reading this book will be interested to know that there is a section on Bill of Materials and Spare Parts. This is a greatly untapped topic and one that I personally believe we need to get smarter on and quickly.

Production has a key role to play in asset management, so we will be spending a great deal of time exploring exactly what is needed there.

As you can see from just this short teaser, there is much to discuss on the topic of asset management, and the intention of this book, and specifically the format of this book, is to drive a larger strategy on how we are going to work together in our facilities to make all this happen. I believe the layout of the book and key learning ideas will contribute to your success.

I need to make a very significant remark considering the remainder of this book and the manipulation of physical assets. For this book, I will be referring to the assets of a company that help to generate revenue. For manufacturing plants this is very straightforward; I'm talking about the production equipment. For facilities and services, I'm talking about the assets that allow a facility to be of use (roof, walls, doors, boilers, chillers, etc.) and possibly any vehicles or transportation equipment. We will discuss the term *organizational objectives* later in this text. One of your organizational objectives might be to be compliant with EPA laws and be good stewards of the environment. In that effort, there are many physical assets that would be used to fulfill that objective. I really just want this book to focus on the assets that allow for revenue generation. I believe that each organization has, as a core objective, to be profitable. That will be a focus of this book. The asset management processes in this book also adapt to the non-revenue generating assets as well (sprinkler systems, backflow preventer, etc.), so please use them. I really wanted to call out this significant element because I wanted you, the reader, to know that business isn't just about making money and I certainly don't mean to imply that it is.

Chapter Endings

Every chapter in this book will end with a review of the chapter's main points. These synopses will form segments of your overarching approach to asset management located at the end of this book. Depending on the particular dynamics and circumstances, some of these chapter abstracts will fit, in the end, to outline a strategy for a commitment to asset management at your location.

In the end, you will be able to review, edit, compile, and pull together the thoughts and ideas to form a very obtainable vision for the future, and make a conclusive argument, on any merit, for a purposeful drive forward. Promise me that you will have fun reading and working through this book, because I intend on having fun while I write it.

ONE

What Are Assets?

"...Language differences frequently—if not usually—erect a barrier to understanding."
—Ed McMinn, *Oklahoma State Daily Devotions for Die-Hard Fans* (p. 17)

Imagine that you have woken from a coma in a familiar land, but communicating with those around you seems to be difficult. Difficult because, although the words are all familiar, the context, structure, and inflection of the inhabitant's speech doesn't conform to your known norms or your familiarity with the spoken language. You can clearly hear every single word, and you know what each word means, but as a structured sentence, your compatriots are not making any sense. How frustrating would that be?

When I was in the military I made an accidental discovery that I went on to prove many times. I had noticed that the higher grade a senior officer (major through general), the more ambiguous and unclear his or her communication seemed to be. I left many colonel and general's offices asking, "What did they just say?" It was difficult sometimes to differentiate between a thought, a direct order, and just an idea. Fortunately, the good senior NCOs (non-commissioned officers) I worked with could help me decipher the intent.

I contend this is the level to which our business vernacular and communication has declined. We are in an organization with other professionals, presumably on the same mission, shooting for the same goals, but for some reason we can't understand our cohorts and they don't seem to understand us. We are not speaking the same language. "When it comes to the language of equipment performance and availability, there doesn't seem to be a common language at all." (Ross, p. xx) And further, "A common language is foundationally needed for a working relationship. In order for maintenance, engineering, production, purchasing, corporate leadership, and others to work and thrive together, understanding the most primary terms is required." (Ross, p. xx)

In the introductory chapter, I mentioned that you would be actively participating as you read through this book. I went on to explain that I was writing a book that had the feel of you and me sitting across the dining room table, drinking coffee and just talking about stuff. Here is your first task, and one that I hope successfully demonstrates this need for us to get alignment on our words, and more specifically, the meaning of those words. Please fill in *your* definition or description of what the following words or phrases mean.

Maintainability:
Reliability:
Asset:
Asset Management:
Asset Management Plan:
Strategic Asset Management Plan:
Maintenance Activities:

Of course there are canned and widely accepted definitions for these words and phrases, but I feel it's important that you and your organization accept an explanation that is organic to your location and your situation. If we all subscribe to Webster's definition, then we might falsely saddle our constituents with a term that doesn't actually resonate with them.

Here is an example of taking a higher authority's definition and convincing others that it means something that just doesn't settle with the population at large. In 2012, the United States government set the poverty line for a typical family of four at $23,050. Of course establishing a standard like this across such a vast and varied area as the United States is a bit of a non-starter. I would agree that a family of four might eke by in Slapout, OK on $23,051 (one dollar above the poverty line). But in New York City, at $23,051, that family is most likely still poor.

I encourage you, before you move on in this book, to completely define the words and phrases previously listed. It is important that you do this to gain initial experience with the interactive design of this book and to make sure that when you say 'asset,' you have clearly stated what you are talking about and there is no ambiguity.

Now that you have completed the assignment, ask two other people in your organization to document their interpretation of the words and phrases. This is important to gauge the degree to which our understanding of simple business-level words differ between members in the same organization. I would expect the answers to be similar, but not exact. No big deal right? I really don't know. But, according to the government, you're poor if you only make $23,050, and thus you qualify for all kinds of benefits. But if you make $23,051, then it's "good luck out there." So, do small differences matter?

Record your two colleagues' responses here:

Maintainability:
Reliability:
Asset:
Asset Management:
Asset Management Plan:
Strategic Asset Management Plan:
Maintenance Activities:

Maintainability:
Reliability:
Asset:
Asset Management:
Asset Management Plan:
Strategic Asset Management Plan:
Maintenance Activities:

Is interpretation and understanding important in the workplace in a reliability setting? Put aside the definition of 'poverty' previously mentioned and in its place add the understanding of reliability, or better yet, a reliability strategy or approach. See Figure 1-1 to visualize how just three slightly different interpretations of 'reliability' and then three additional thoughts on connecting issues can lead to an unmanageable spare parts matrix.

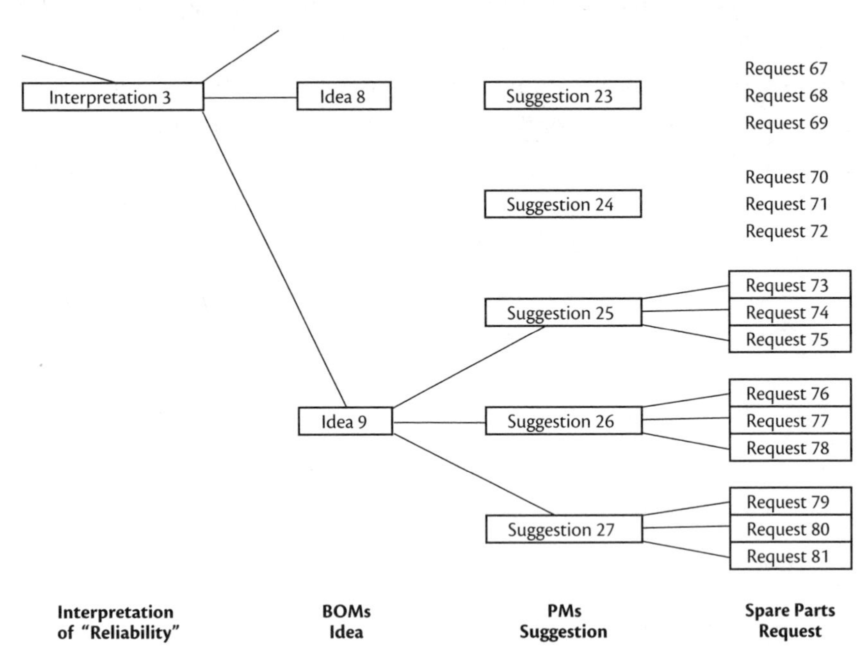

Figure 1-1 A portion of the results for a slightly different interpretation of 'reliability'

Figure 1-1 is highly unscientific but shows what I believe to be an all too real example of how we can end up off the mark by starting off with a loose definition. The figure shown is a portion of the larger results that stem from asking three people to define what reliability means. Then, three additional people were asked to take those three definitions and describe their ideas for the Bill of Materials, and so forth until the group ended up at eighty-one unique requests for a spare parts package to support the exact same machine.

Has it been your experience that you keep adding spare parts to the storeroom for equipment that has been in your plant for over ten years (or longer)? This phenomenon might be the result of different people with differing ideas of what is needed to support an operational asset. It is certainly a result of not having a clear process for determining how to put together a spare parts package. The skill of putting together such a package requires some decision-making criteria. This will be discussed later, in Chapter 4.

To summarize this introduction to Chapter 1, we have to understand each other in order to work, live, play, and prosper amongst one another. In describing the day of Pentecost, St. Luke painted the scene of a large crowd, made up of religious people from every country in the world: "They were all excited, because each one of them heard the believers speaking in his or her own language." (Acts 2, 6-7 GNB)

Imagine the transformational feel of having everyone throughout your organization speaking in a language everyone can understand. A clear, crisp message can only be attained if the intended audience is part of the process of communication. We can start by first understanding exactly what makes an asset an asset in our company.

What Makes an Asset an Asset?

Just a few paragraphs before, you were asked to record your definition of the word 'asset.' At the risk of seeming like busy work, please document a combined definition, using your previous thoughts with those of your two colleagues. Please record that definition here:

Asset:

For the remainder of this book, or until you find it necessary to change this melded thought, let's assume that this is your organization's working definition of what an *asset* is. This exercise is important because now we are going to discover what makes an asset an asset in your company.

The working definition I use to explain what an asset is can easily be understood to be "*any tangible or intangible item, thought, process, concept, or substance of value that can be manipulated for financial gain.*" Ok, that got wordy, but I wanted to incorporate a lot of ideas into that working definition. In the business world, anything that has value and can be used for fiscal gain is an asset. The opposite of an asset in business is a liability. The two (assets and liabilities) make up what is known as a Balance Sheet. When a board member or corporate executive asks, "On the balance, how are we looking?" they are asking what the value difference is between assets and liabilities. As you can imagine, having more liabilities than assets is one attribute of bankruptcy.

This is not an exhaustive list, but rather an example of what an asset might look like in your organization. Very quickly, here are some assets you might consider:

- Capital assets
- Human resource assets
- Financial assets
- Property assets
- Trade secrets or proprietary assets
- Processes
- Inventory
- Accounts receivable
- Your company's brand

A careful review of the specific definitions of each of these might help in understanding their intrinsic value and to grow an intuitive desire to properly manage these assets. I caution you, these interpretations are not dictionary definitions, but rather my definitions based on three decades of field research. I strongly suggest that as you read this, you give some thought to how these assets are interpreted in your organization.

Capital Assets

I refer to capital assets as the 'stuff that makes the stuff.' This may be one of the easiest concepts to grasp. Capital assets may be a piece of equipment or machinery (e.g., office machinery, conveyor, oven, vehicle, etc.) that your company has 'capitalized,' indicating that you bought it to enhance, grow, or sustain your business for which the federal government rewards you by allowing you to depreciate its value over time. Depreciation amounts and time frames are dependent upon what type of asset it is, its cost, and use. Capital assets are usually identified in a business by having an individual asset tag and asset number. It is not unusual for maintenance departments to get involved by performing an annual 'asset audit' to assure the accounting department that all the assets being 'depreciated' are in fact still on the premises.

Several years ago, some crafty accounting loopholes were discovered that allowed for what were traditionally expensed repair activities to be 'capitalized.' This seemed like fuzzy math, but it also appears to be legit. The unintended consequence is that maintenance and the storeroom have to track spare parts as capital assets. Very few companies or organizations do this very well. If you track these new capital additions and you're good at it, then consider yourselves leaders of the pack.

There will be greater discussion on capital assets as this is actually the focus of this book. For now we'll keep this as just an introductory session.

Earlier I shared my working definition for an asset as, "*any tangible or intangible item, thought, process, concept, or substance of value that can be manipulated for financial gain.*" In Table 1-1, please list a single capital asset in your facility and at least two ways this asset can be (or is) manipulated for financial gain.

Table 1-1 Methods for Manipulating One Asset for Financial Gain

List one capital asset:	
How is it manipulated for financial gain, means #1:	
How is it manipulated for financial gain, means #2:	

As an example of what I was looking for in Table 1-1, my fictitious company has a mechanical press. There are many ways my company manipulates this asset for financial gain.

One manner in which they do this is to run the unit 24/7. Secondly, to speed up changeovers, my company has modified the press for quick changeovers to increase its overall availability.

Look back at your entry on Table 1-1 and confirm that your company uses assets for financial gain. This is not a negative attribute. That's what companies do. They manipulate assets to make money. If it helps, you might consider using the term 'use' instead of 'manipulate.'

Now, the harder exercise: if the asset you listed is truly an asset and valuable, in Table 1-2 please indicate the manner in which your company manifests (or demonstrates) the belief that this particular asset is of great importance to the financial success of the organization. For example: my company has awesome preventive maintenance, clear and complete spare parts availability, and solid drawings and schematics for this mechanical press.

Table 1-2 Demonstration of an Asset's Importance

List one capital asset:	
Demonstration of asset's importance #1:	
Demonstration of asset's importance #2:	

Look back at Table 1-2. Can you honestly convince people that your company values its assets, and actively demonstrates the importance of the equipment's role in accomplishing the mission of the business?

Human Resource Assets

Years ago, I was teaching a class in the middle of the United States and we were just in the 'introduce yourselves' phase. A student in the center of the back row was wearing a large, red sweatshirt with a giant letter "N" on it. I asked him if he was a Cornhusker's fan. He proudly pointed to the letter on his shirt and proclaimed, "Yep, where the 'N' stands for knowledge!" I instantly figured him for an OU man (Author's note: I graduated from Oklahoma State University).

Aside from being a funny story, the gentleman made a poignant observation. It is the knowledge that is important. When companies or individuals tell me that people are their greatest asset, I usually respond, "No they're not. It's the knowledge and skills that they possess that make them an asset." Let's face it, a new employee is not an asset on day one. Heck, they're probably a liability. I jokingly tell my classes, "Jerry, stand over there and don't touch anything." A human resource asset isn't truly an asset until they know something and can do something that leads to a fiscal gain for the company.

To truly reflect the value of people as assets, a human asset, there needs to be a comprehensive training and development program built around the person and the position. For example, Figure 1-2 might reflect the different roles of a maintenance planner/scheduler in your organization.

Planner/Scheduler
Planner
Scheduler
Facilitator
Maintain BOMs
CMMS gatekeeper
PM/PdM overseer
Picklist generator
Data analyzer
Report generator
KPI/Metric generator

Figure 1-2 Roles of a Planner/Scheduler

If Figure 1-2 reflects the roles that a planner/scheduler might fill at your facility, there should be an objective-based training protocol for each. Take, as an example, what a training matrix might look like for the planner/scheduler for the role of PM/PdM (Preventive Maintenance/Predictive Maintenance) overseer. Figure 1-3 is an example of this matrix.

Figure 1-3 shows the start of a training matrix for the planner/scheduler. The first few columns of this particular matrix are meant to show some skills that are necessary for the planner/scheduler's role as the overseer of the PM/PdM program. Note that the legend for how to read this matrix is in the upper left-hand corner.

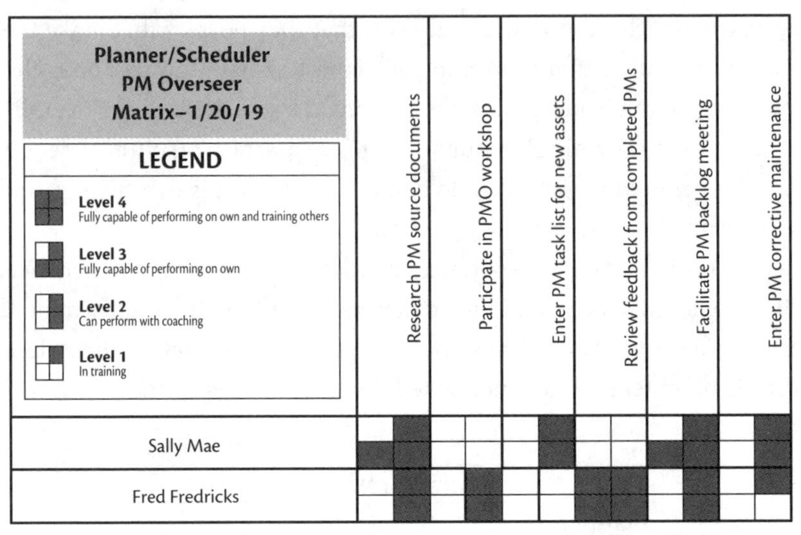

Figure 1-3 A training matrix for the Planner/Scheduler role of PM/PdM overseer
Adapted with permission by Marshall Institute, Inc.

To demonstrate the lengths to which your organization will ensure that a human resource is actually an asset (contributing to the fiscal benefit), in Table 1-3 indicate a single individual's first name (in your company), one role they fulfill, and two skills they are required to have as part of their transformation into a greater asset for the company.

Table 1-3 Increasing the Value of an Associate
Adapted with permission by Marshall Institute, Inc.

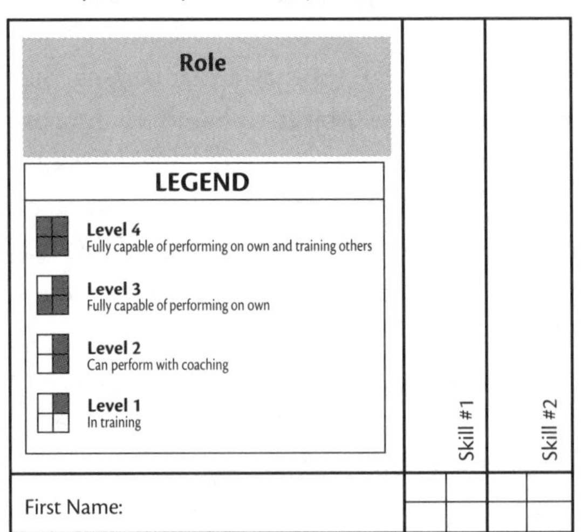

Just to recap, for human resource assets to truly be assets, they need to know how to do something that contributes to the fiscal prosperity of the company.

Financial Assets

Financial assets might be the easiest and most relatable concept for us as individuals to comprehend. Why? Because we have so much personal history with this type of asset. Financial assets are the answer to, "How much money do you have, and what are your investments and your level of liquidity?"

I personally have four bank accounts, two business checking accounts, one personal checking, and one personal savings account. Years ago, I was in the bank transferring some money between accounts and getting some cash and the teller asked me a strikingly personal question. "Mr. Ross," she said, "why do you have four bank accounts?" I responded that I needed four accounts because the bank only insured each account up to $250,000. She looked amazed and asked quite sincerely, "You have one and a half million dollars in our bank?" I did not correct her math, but I did ask her to count back my cash, twice!

The financial assets are basically the portfolio that is the cumulative accounting (pun intended) of all your money and investments. I won't ask you to list your financial portfolio. But, understand that businesses depend on cash flow. Companies don't like to, nor do they want to sit on large sums of cash; however, they need to have some level of cash reserves. The financial assets are often tied up in capital assets, and investments into other businesses.

Property Assets

I've done a considerable amount of facilities consulting in which the client is not a manufacturing operation, but rather a service or information organization. Not unlike your own location, their property asset value is dependent on the 'real' property that they own. It would not be unusual for companies to lease space rather than own it outright, or buy on terms.

Sometimes, and very interestingly, a property asset can become a liability. This can occur when the property the organization owns becomes unusable, dangerous, or just not desirable. Think of a location that is a superfund cleanup site. This would clearly be a liability for a company.

For property to continue to be an asset, it has to be cared for in such a manner that it maintains its value for service. It would not be beyond the pale for the land to be more valuable than any buildings that sit on it.

This might be urban legend, but in an interview, Ray Kroc (McDonald's) was asked how he felt about being the hamburger king of America. He corrected the interviewer to say that he wasn't the *hamburger* king, but he was the *real estate* king of the United States. He owned all the land under all the McDonald's restaurants.

Trade Secrets and Proprietary Assets

The intellectual property of your company has a real value. Imagine the trade secrets and secret recipes that have built the enterprise that now employs you. Patents and trademarks have a shelf life, and companies work quickly to capitalize on the market and make a buck.

Trade secrets transcend the discussion of assets and work their way into the maintenance and safety realm through OSHA's 29 CFR 1910.119, Process Safety Management. Trade secrets are one of the fourteen elements listed and explained in this federal regulation.

Not necessarily a trade secret, but as an interesting aside, I was awarded a patent in 2018 after having submitted the paperwork eight years prior. Look it up, patent #9,636,832; apparatus and method for spirally slicing meat. I made exactly one dollar off of it. It might be time to open a fifth bank account!

Inventory

Now we're talking! Inventory can be a tremendously valuable asset, but it can also be a heartbreaking liability.

There are five major inventory categories in most operations:

1. Principal supplies
2. Work in Progress (WIP)
3. Finished goods
4. MRO (Maintenance Repair Operations or spare parts)
5. Office supplies

Principal supplies are the raw materials that we use to make, package, and ship our product. Work in Progress is self-evident and is literally the work that is on the floor in the process of being 'made.' Finished goods are the products that are packaged, and ready to go to the customer. MRO are the spare machinery parts. It may be that your company includes consumables in this category as well. Office supplies are the actual administrative supplies needed to keep the office and staff functions running.

The most highly coveted and most secure inventory is without a doubt the office supplies. If you need a box of staples, you usually have to see "Marge" up front and she will ask what you need an entire box of staples for. She will most likely give you one sleeve of staples and snap that one in half in front of you.

That might have been said in jest, but keep this in mind. Of all these types of inventory, the *right* inventory is an asset, *mismanaged* inventory is an expense, and the *wrong* inventory is a liability.

Years ago I was the general foreman for a mini-steel mill in the Tulsa, Oklahoma area. My boss, the gentleman that ran the melt shop, told me that we (the plant) made three thousand grades of steel at that plant. Get your head around that number. Our mini-mill made three thousand different chemical compositions of steel. Three thousand, my boss said. And some of those on purpose! My boss went on to tell me that we only sold ten grades of steel at that location, but we had produced 2,990 off-spec grades of steel. Most of that was 'out back' being chopped up by a contractor for us to re-melt again. That is an example of an inventory that is a definite liability.

Accounts Receivable

When putting a valuation on a business, one of the elements you might consider is the business they have on the books. This would include contracted work, possibly proposals out for work (but not typically), but would definitely include the outstanding accounts receivable.

A small business owner once described a client to me like this, "They are our best client. They owe us two million dollars." I suggested that he find a different definition for 'best client.' This is a tentative measure for sure, but it could be argued that money projected to come in is in fact a kind of asset.

Your Company Brand

Consider how much your company brand and branding have to do with the overall fiscal success of your company. What's in a name? Turns out it is everything. A name, a signature, or easily recognized logo can set a company head and shoulders above its competition.

Early in my consulting work I had occasion to speak with a company executive who was new to the organization I was consulting with. He had recently been hired away from Nike. For years he had been an executive vice president. He asked if I could name any of the top three most recognized brands in the world. I thought this was a trick question, and answered, "Nike?" Nope. "McDonald's?" Nope. "Starbucks?" Nope and nope. Not even close. Amazon, Facebook, and Twitter. Turns out, as he remarked, the top three most recognized brands in the world don't make anything. That should be a sobering thought to everyone.

The following letter appeared in the January 2019 in-flight magazine, *Southwest: The Magazine* (Southwest Airlines' [SWA] magazine). This heartwarming story is from the section devoted to highlighting SWA associates performing great customer service. I want you to notice how the story, sincere to be sure, helps to bolster the brand that Southwest Airlines wants to keep strong as they market to business and family travelers with a touch of humanity:

> I always get a little nervous traveling with my 5-year-old daughter, Mikaela, who has autism. She's been doing much better on airplanes, but I always ask to preboard because walking onto a full flight can be overwhelming. I did this for my recent flight out of Burbank, California,

> and was helped by Customer Service Agent Christopher Ulrich. As we were boarding, he quickly handed me a little booklet that I assumed was something for my daughter to draw on during the flight. After we settled in our seats, I realized it was not just a plain booklet, but the most amazing present I've ever received from a stranger. He had illustrated a story called "Mikaela's Flight." Needless to say, I cried tears of joy the entire flight. Part of my daily struggle is never knowing what may cause my child to break down. To know that someone cared and understood really put me at ease. Christopher's actions perfectly illustrate why Southwest considers their People [*sic*] their "single greatest strength."
>
> —Brenda Yeh, Southwest Customer

A company's brand takes a long time to become iconic but can be lost in an instant by bad or corrupt activities. The brand is an asset that must truly be nurtured and protected.

It would be reasonable to ask why so much time and work was devoted at the start of this book on asset management to lay out some groundwork on common forms of 'assets.' It is important for the purpose of developing an intuitive desire to properly manage assets. Take, for example, the last section on the Southwest Airlines agent, Christopher Ulrich. Did the personal touch that Mr. Ulrich displayed on that flight indicate that SWA is a company that values its brand and its associates? So much so that it's almost as if their associates *are* their brand.

How Do Assets Contribute to the Success of the Company?

Taking into account the conversation we just had regarding the different kinds of assets, let's create some specificity and begin to tailor our focus on the capital assets of business. There are a multitude of books and resources on financial success, property management, and human resource advancements. But the core idea behind the asset management movement is the attention given to the *capital assets*, or as defined earlier, "the stuff that makes the stuff."

In his perennial best seller *Rich Dad, Poor Dad*, Robert Kiyosaki made his *rule one* very clear: "You must know the difference between an asset and a liability, and buy assets." (p. 58)

Mr. Kiyosaki's book is not a maintenance or reliability book, but it perfectly illustrates the relationship between assets versus liabilities; and income versus expenses. Consider the next series of figures shown to animate the flow of 'money' through an organization, your organization.

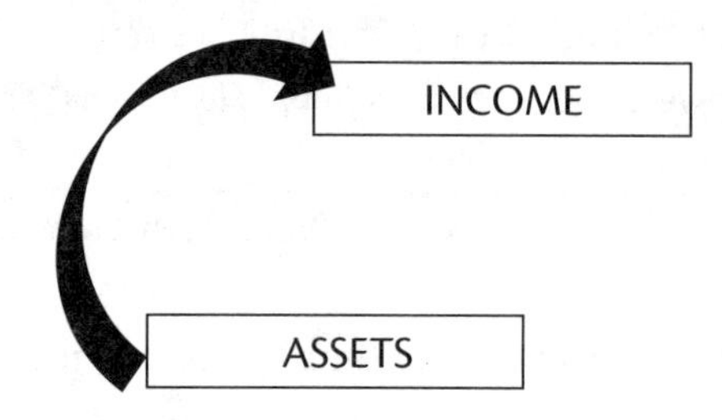

Figure 1-4 Assets make income

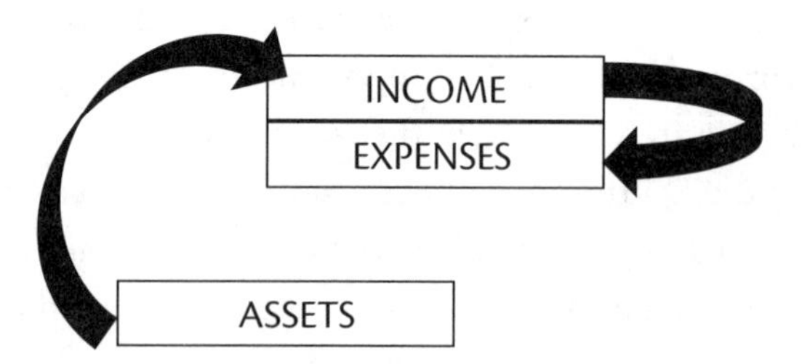

Figure 1-5 Income is used to pay the expenses

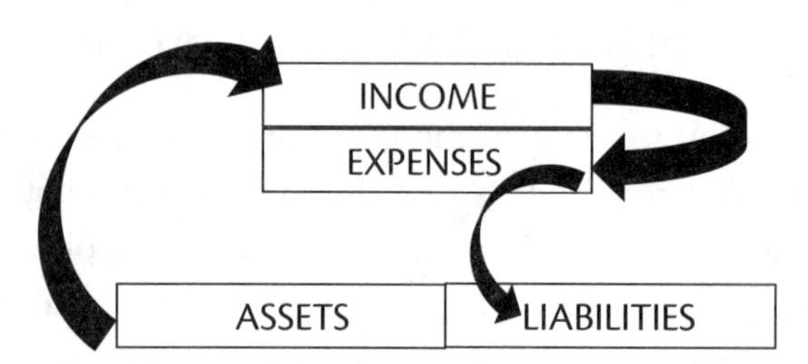

Figure 1-6 After expenses, liabilities are serviced

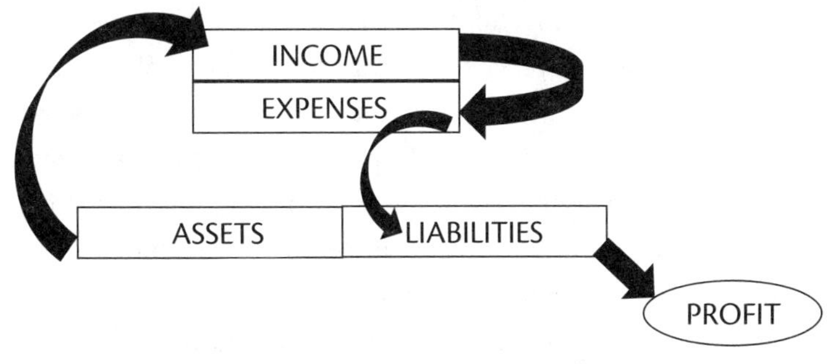

Figure 1-7 The result is profit

Using information gathered from Figures 1-4 through 1-7, complete the following sentences:

Only	make income
Companies take that income and pay their	
After that is paid, they service their	
The result is	

I'm extending an author's prerogative and letting you know that the answer to the last fill-in-the-blank sentence is 'profit.' Companies take that profit and reinvest it back into the company. They buy more assets (to quote Kiyosaki, "buy assets") to make more income. Remember this: *only assets make income*. Highlight, circle, and cast that last sentence in stone.

Aside from the clear circular reasoning that it is only the assets that can contribute to the fiscal health of the company, it bears mentioning that rolled into the definition of the 'success of a company' is the idea that it is through the assets that we also move in the direction of satisfying the corporate goals and objectives. The presumption is that some of these high ideas are fiscal in nature.

Keep that in mind and consider that in the ISO 55000 standards, an asset is any item that has potential or actual value to an organization. The value will vary based on the organization and, what you will come to appreciate later, the organization's stakeholders. An asset doesn't necessarily have to be tangible or financial in order to have value or potential value.

A colleague of mine once told me that in his opinion everyone in an organization owes their paycheck to the smooth and reliable operation of the capital equipment of the enterprise. This interpretation lends itself to all facets of industry. A significantly important asset in your company, in some small way, determines if you get paid or not, and oftentimes how much. Extend that idea out and you can see that the operation and continued operation of our plant and facility assets determine where we live (socio-economic), what kind of car we drive, and even where our kids go to college. That is or should be a powerful thought.

I borrowed this idea, and a central tenet of Total Productive Maintenance (TPM), to make this observation in my first book: "Everyone in the organization has, as a responsibility for their role, an understanding that their work should be profitable to the company. On the subject of equipment reliability, everyone has either a tacit or implicit responsibility to make sure the operation runs smoothly and reliably. When capital equipment is involved, everyone is responsible for equipment reliability." (Ross, p. 49)

Let's put that idea to the test. In Table 1-4, list the same plant or facility asset you listed in Table 1-2. Get personal now and list one way that you personally ensure that the listed equipment runs reliably. Finish out Table 1-4 by recording how the company benefits from the reliable service of this asset and two examples of the negative effects your fellow associates might suffer if the asset was not running reliably. An example of that last request might be that associates are temporarily laid off without pay while the plant is temporarily shut down.

Table 1-4 Our Role and the Benefit of Reliability

List one capital asset:	
How do *you* ensure reliability?	
How does the company benefit from this asset?	
First way an associate suffers:	
Second way an associate suffers:	

This idea of being connected to the asset through a shared responsibility is the crux of the asset management concept. Maintenance can't do it alone, and production and other agencies need to be other cogs in the wheel of accountability.

There was a party game that ran through the United States some time ago called "Six Degrees of Separation." The idea was that everyone in the U.S. was connected to the actor Kevin Bacon through a series of networking by no more than six degrees, or nodes.

For example, I know Bill, who has a brother in Los Angeles who is a police officer. Bill's brother Rick worked on a crowd control assignment in downtown L.A. on a film shoot. The actor on set that day was Robert Sedgwick, who is the brother of actress Kyra Sedgwick. Kyra Sedgwick is married to Kevin Bacon.

Bill-Rick-Robert-Kyra-Kevin. Five nodes or degrees.

I was at a major Fortune 50 company in the Northwest years ago teaching a strategic maintenance class and for lunch we all went to the cafeteria. We were at world headquarters, so the spread was quite impressive. I was in the hot entrée line and noticed that the young man serving the main entrée was named (you guessed it) Kevin Bacon. My response was of course, "no way." Way!

I asked if I could take a picture with him and send it to my family. Now, thanks to this chance encounter, my entire family is one or two degrees away from (a) Kevin Bacon. Ok, not *the* KB.

If in some crazy, arbitrary way, we are all connected to someone as random as Kevin Bacon by no more than six degrees, can't we somehow also be purposefully connected to the reliability of the very 'thing' that generates our paycheck?

We *are* connected to our equipment and we have real reason to have great expectations from the continued service of our machinery.

Assets are expected to perform at a predictable level to ensure a predictable return. In the development of our capital assets, there is an expectation for a return on investment, or ROI. It is very likely that through some acceptable financial and production calculations a company can estimate with a degree of precision exactly how long it will take to make its money back on new equipment purchases. These calculations include assumed levels of productivity and also should account for operating and maintenance costs. These calculations result in an *ideal* value and may include some contingency planning.

The projected value of an asset to the company, for its ability to produce the product or service, is primarily how an asset contributes to the success of a company. The proper operation of an asset, not only now but in the future, is what quite frankly pays all the paychecks in your organization. Does it seem to you that people know that?

Where Do Assets Come From?

Keep in mind that it's through the successful manipulation and utilization of the assets that a company makes money. Undoubtedly, your company has many lofty and laudable objectives and mission goals, but making money for its stakeholders and shareholders has got to be at the top of the list. This theme has to be central to our understanding of the true value that assets have to the success of an organization's objectives and mission. This not only applies to capital assets, but all forms of assets. It is the capital assets that require our focus in this book.

So, where do assets come from? In the space below, please record all the different means of acquiring a capital asset that you have experienced.

Here is my short list that might match yours:

- Another plant within the company
- Another plant from another company
- From surplus sales, rebuilt or as-is
- New from an OEM (Original Equipment Manufacturer)
- Modified version of another existing machine
- Made in our own shop

Assets, it seems, can come from many sources, but there are some qualifications that need to be met before they are truly assets. Give this some consideration. What details or information do you feel you should have or have determined before you commit to purchasing an asset or modifying an existing asset? If assets make income, and they are the only things in business and life that do make income, what are some qualifiers? List your thoughts here:

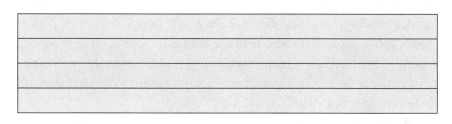

The objective of this simple exercise is to try and connect an asset purchase to an asset purpose.

For instance, if I were in the trucking business, I might consider such details as:

- Hauling capacity
- Distance per tank of fuel
- Fuel mileage
- Turning radius
- Height of the vehicle
- Ease of operation and maintenance
- Standardization with the rest of the fleet
- Annual operating costs
- Insurance and taxes

Here is a scenario to consider: I'm in the market for a new truck to add to my trucking fleet for my trucking business. The list just provided are the attributes that are most important to my purchase decision. In order for my new truck to contribute financially to the plan that I have for it, these are the characteristics that I need to look for, ensure, and guard. I do this to generate the most *income* from my *asset*.

This section is titled "Where Do Assets Come From?" We started off with a listing of possible places that an asset can physically come from. How often do our aspirations for a new asset (new or used) fail to pan out because of the compromises we make along the way when landing the new asset?

If we buy a used or surplus piece of equipment, what are the odds that the new (new to us) asset will live up to the reason we are buying it in the first place? In the truck purchase example I gave earlier, for the categories I listed, it would seem reasonable that I have real interest in specific performance deliverables. It is unlikely that a used truck would hit all the gates, exactly as I've listed them in my concerns. Yet I still believe this 'new' vehicle is going to deliver as I hoped; essentially, becoming the asset that will make the projected income that was anticipated.

Where assets come from is as critical to establishing a reasonable performance and income curve as it is practical to understand that any compromises and alterations on the 'needs' list of an asset will absolutely also alter the 'deliverables' list.

Has it ever been your experience that your company bought a used asset and expected the maintenance department to fix it up to use for all-out production? How did that work out? There is nothing wrong with used equipment, but just keep in mind that a rush for the fiscal sensibility of buying used equipment doesn't always lend itself to dovetailing perfectly into our production model. Neither does buying an off-the-shelf new item either. This is a very important dance that is done to make sure we get the right asset for the purpose.

If we are going to manage assets to make income it would make sense that we go into the proposal with the greatest level of confidence in pulling that off, if for no other reason than for the longevity of the production effort. Asset availability is equally important to asset capability.

Darrin Wikoff mentions in his work, *Leader's Guide to ISO 55001: Asset Management System Requirements*, that "Availability is the result of the asset base being both Reliable—the probability of conforming to the desired function without failure—and Maintainable—the ability of your management system to restore functionality after a non-conformance has occurred." (p. 9)

Regardless of the origin of the assets in your facility, they need to *start* their history at your location fulfilling the purpose they were purchased to fulfill. We manage assets to make income. If they weren't designed to perform the work in the first place, it is likely that we will have a history of frustration and disappointment.

But who is responsible for ensuring asset capability in the first place?

Who Pulls The Trigger?

Who determines what asset to purchase and when to actually get it? In your organization, who is chiefly responsible for making asset decisions? (It could be a group of people.) Write that source here:

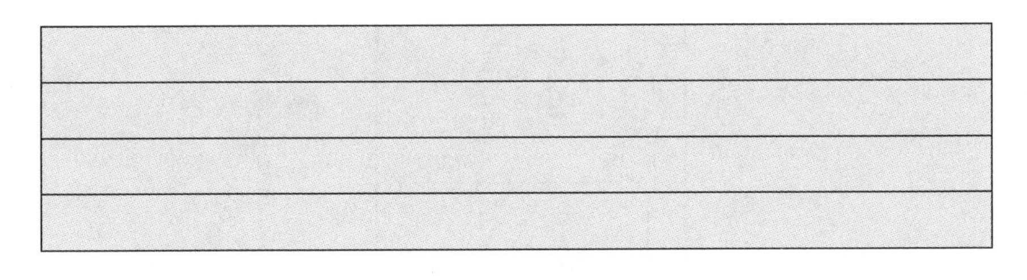

In business, asset purchase considerations are often, if not always, determined based on a projected benefit to the company. These purchases will certainly include facility assets such as boilers, air compressors, and even plant roofs. In manufacturing and facilities industries we want to focus on capital assets that actually make our product or allow us to provide a service. Note: if you are in the facilities maintenance position, of course the facility assets themselves are the target of our study.

Regardless, these purchases are often made as part of a projected business case that was conceived well in advance. Couple that with the lead time required to design, build, and install assets and it's apparent that any project must have the timing of a moon landing to fully capitalize on the gains that were projected much earlier. Figure 1-8 is meant to show this intercept between planning for a capital asset and actually gaining value from the asset decision.

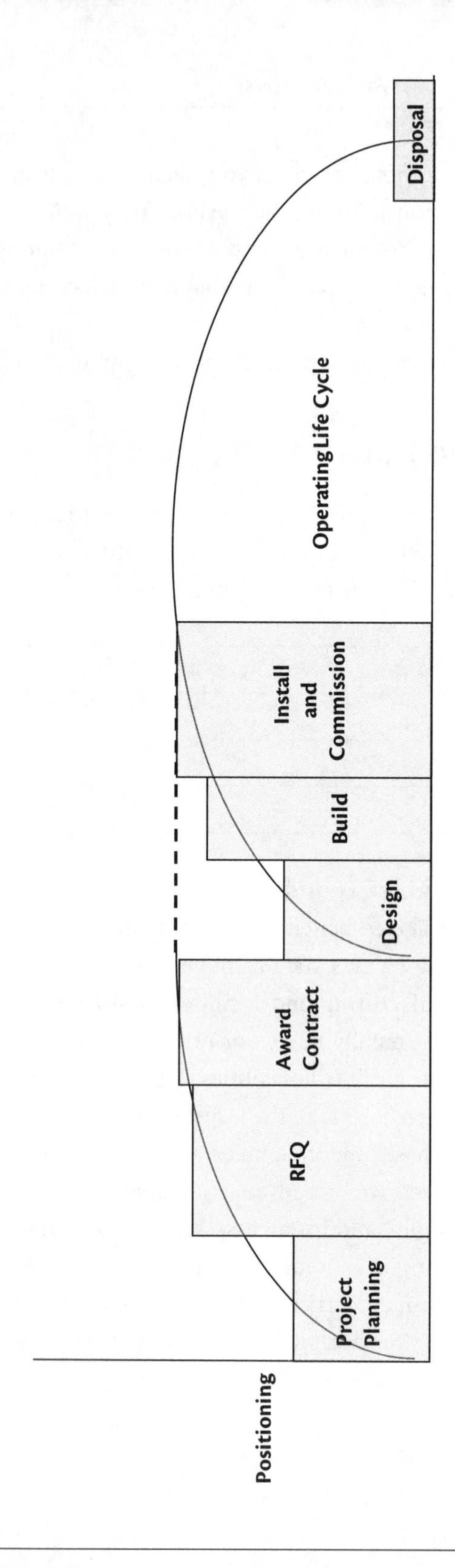

Figure 1-8 From project planning to asset operation to disposal

The '*who makes the decision of what assets to purchase*' is pertinent to an asset management discussion because it begs the question, "Are you more engaged with an asset's care and use if you are intimately responsible for it?" ISO 55000 directs those adhering to this standard that an organization's top management, *employees*, and stakeholders are the groups responsible for conceiving and executing what is referred to as "control activities." These activities might include: policies, procedures, and performance measuring and monitoring techniques. This is done to identify and capitalize on opportunities and to reduce risks. Interestingly, ISO 55000, in its direction to reduce risk, indicates that risk should be lowered to an "acceptable level." It would not be uncommon in our factories and facilities around the world to be very light on an exact understanding of an *acceptable level* of risk.

Ron Moore, in *Making Common Sense Common Practice*, put it this way when describing a scenario of payback analysis on capital equipment: "If we could reduce those production rate losses and minimize the maintenance costs by taking the advice of the people who are in a reasonable position to know what the problems are, then perhaps we could reduce future costs or increase future benefits." (p. 127)

The '*who makes capital asset decisions*' should be a team that is fully engaged from concept to operation. This team should be cross-functional and made up of the usual suspects:

- Engineer
- Maintenance leadership
- Production leadership
- Maintenance hourly
- Production hourly
- Safety
- Purchasing
- Storeroom

The charter for this team is simple. Take into consideration the last section on *where do assets come from*, and factor in that a small but effective team as previously described can have a huge impact on covering all the bases in designing equipment for the purpose intended. It is likely more possible that a team with this makeup might have a better chance of achieving a design and fielding an asset that is successful. Conversely, give thought to an asset that is completely conceived and scoped by headquarters or an engineering corps that is closed off in the corner office. Which scenario would net a better result in the long run?

Try your hand at putting some names behind the titles. In a reprint of the bullets listed previously, jot down the names of people at your facility that might be of real value to fill a position on the next capital asset design team.

Engineer:
Maintenance leadership:
Production leadership:
Maintenance hourly:
Production hourly:
Safety:
Purchasing:
Storeroom:

On the subject of *when* assets should be purchased the answer would naturally have an element of company business plan exposure. Capital equipment is usually capital intensive, meaning that it is costly to purchase assets. A cost-benefit analysis is typically needed at a bare minimum to assess if a capital asset and its operation will gain the fiscal benefits being projected. Time is usually a critical component when calculating the lead times needed to secure the necessary project funding and get the equipment built and installed. This coordination requires a solid grasp of market trends and financial projections way out into the future.

We won't even touch on securing principal supplies, packaging, and other requirements that might come from a new product being run, or an increase in production of current SKUs (Stock Keeping Units). There is so much to factor into an asset purchase that it would not be uncommon for organizations to feel that they don't have time to spin up a team for every asset purchase being considered. Yet we always have time to overcome the reliability and maintainability issues that are inherent in whatever the folks upstairs come up with. There was an old Fram oil filter commercial in the 1970s that had the famous tagline, "Pay me now or pay me later."

If we've heeded such advice as given over the last few sections, we would now have an asset that is truly performing to the levels we intended. This is an asset that we can and should manage for the continued success of our company. But, how long should we keep this asset?

When Enough Is Enough

When the conversation turns to asset life, or even calculating life cycle costs (LCC), the ISO 55000 standard introduces an interesting twist. We learn that an asset's life does not necessarily end when the machine is no longer of use for *our* purposes. Instead, the asset can provide value to more than one organization over its (the asset's) lifetime.

We've actually seen this before. Anyone who has ever cared for an automobile at a very high level to protect the trade-in value has an absolute appreciation for the sentiment the ISO standard just referenced. Here is a story demonstrating this point that fits perfectly into the discussion on when to dispose of an asset.

My first assignment in the Air Force after the Aircraft Maintenance Officer Course was to Wurtsmith AFB, MI. I was a flight line maintenance officer (2nd Lieutenant) and I had good friends in the Company Grade Officers Council (Lieutenants to Captains). One of my friends, Terry, was a lieutenant at the base motor pool. The Air Force motor pool was made up mostly of staff cars. This was a time in the military of tight budgets, and as a result, the Air Force had purchased almost exclusively the Dodge K-car platform, and it seemed that the majority of cars were the Reliant K-car.

It's widely known that cars are individually identified by their VIN (Vehicle Identification Number). These cars were no different. Terry told me that each car cost about $8,600 landed at our base, all decked out with the Air Force required package: AM radio, roll-up windows, manual locks, etc. He told me that each vehicle had in essence a 'bank account' that had roughly $10,000 in it, and each time the car was serviced (other than fuel), the labor and material charge was subtracted from that account. Once the car's balance hits $0 (zero), the car was sold.

I'm not confirming the numbers or Terry's accounting of these details, but I remember feeling that this was a very good idea. At what point do we decide to stop spending money on an asset and just buy a new asset? Imagine that we are caring for the assets to ensure the highest trade-in value.

I often tell clients that if we don't have some trigger set in our process to evaluate where we are on spending money towards the upkeep on an asset, we are likely to find ourselves having spent $2.3 million on an asset that costs $300K brand new.

When my son graduated from college he still had the car I bought him while he was in high school. He asked me, "Dad, how do you know when it's time to stop spending money on an old car, and just buy a new car?" I told him, "When *you* can afford it."

I wonder how many companies actually build an asset disposal plan into their life cycle projections, including the funding for such actions. Ramesh Gulati re-prints a value in *Maintenance and Reliability Best Practices*, of <5%. (p. 177). Disposal costs for an asset with an LCC of $1,000,000 should be less than $50,000, as an example.

A good business plan has an exit strategy. What is your capital asset exit strategy?

CHAPTER SUMMARY

It makes sense to start a book on asset management with a short discussion on what makes an asset an asset. I contend that although we call many things 'assets,' in practice we don't treat them like assets. Give some thought to the care and upkeep of your capital equipment and the knowledge and skills of your people. Are you currently providing the necessary attention and direction needed for each, as if the business literally depended upon it? In most cases we are doing what we can, but not executing the care at the highest levels possible.

NOTE TO THE READER: What follows is the start of a compelling business case to adopt a formal asset management plan for your company. I've had success with this format in the past and I know it will work in this case as well. Fill in the spaces below using the responses to this chapter's work. Much of this same material will appear in the back of the book in the closing chapter. It is my hope that the material in the back of the book, once completed, will net a comprehensive and compelling case to encourage your leadership and associates to move forward on this critical journey and provide a blueprint for asset management.

Our Business Case

We often begin our journey in asset management by setting off on the wrong foot. The manifestation of that misstep is not having an acceptable definition for key words and phrases. For the purposes of our discussion on asset management, here are some key words and phrases and the working definitions at this location:

Maintainability:
Reliability:
Asset:
Asset Management:

It must be the focus of this organization to recognize that there are many types of assets, including but not limited to:

- Capital assets
- Human resource assets
- Financial assets
- Property assets
- Trade secrets or proprietary assets
- Processes

- Inventory
- Accounts receivable
- The company's brand

For the extended purpose of asset management, we are going to focus on capital assets, recognizing that the effective deployment and utilization of capital assets make income for the company. Assets, it turns out, are what we manipulate to make income and ultimately a profit. Robert Kiyosaki animated this for us in the following figure.

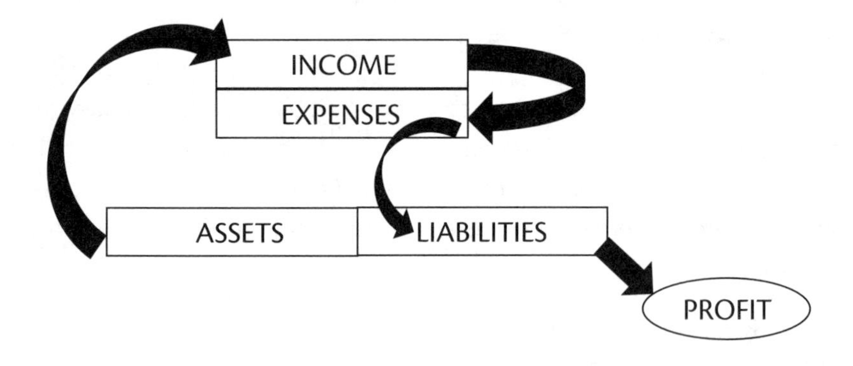

In each case, in order to extract the most income, these assets must be honored and cared for as if the balance of the organization were hanging from their success. Because it is.

Our focus is on capital assets. In a way, each of the members of our company has to see the connection between their work and the proper maintenance and reliability of the company assets (capital equipment). It is with this connection that we all become engaged in *asset management*.

Through this connection to the asset and its purpose of creating income for the enterprise, we have to articulate exactly how this happens. The following is an example of how a single plant asset is manipulated to make income for the company.

List one capital asset:	
How is it manipulated for financial gain, means #1:	
How is it manipulated for financial gain, means #2:	

Further, to demonstrate the importance of the company's assets to the business aim of this organization, here at this location, our care manifests itself as shown in this example:

List one capital asset:	
Demonstration of asset's importance #1:	
Demonstration of asset's importance #2:	

It is strongly believed and virtually dictated in ISO 55000 that the employees be given a voice in equipment design and construction. As a result, we suggest that all design teams have a standard makeup to include the following representation:

- Engineer
- Maintenance leadership
- Production leadership
- Maintenance hourly
- Production hourly
- Safety
- Purchasing
- Storeroom

Turn to the last section of the book and record this same information in the comprehensive strategy section.

TWO

The Asset Management Standards

"...Standardized work was never intended by Toyota to be a management tool to be imposed coercively on the workforce. On the contrary, rather than enforcing rigid standards that can make jobs routine and degrading, standardized work is the basis for empowering workers and innovation in the work place."

—Jeffrey Liker, *The Toyota Way* (p. 145)

We've arrived at the heart of the matter. As laid out in Chapter 1, our focus on asset management will be on capital assets going forward. Therefore, we are going to take a comprehensive look at what the standard refers to as *physical assets*. Physical assets, in the ISO 55000 series of standards, refer to many different categories, but it is the *equipment* that we will be focused on in this chapter, and in this book. To be absolutely clear, the ISO 55000 series of standards refers to many types of assets (e.g., financial, human, etc.), but we are going to concentrate on the physical, or capital, assets. This is essentially what I call "the *stuff* that makes the *stuff*."

If you are not familiar with the ISO 55000 series, there are actually three standards to consider:

1. ISO 55000 Asset management—Overview, principles, and terminology
2. ISO 550001 Asset management—Management systems and requirements
3. ISO 55002 Asset management—Management systems and guidelines for the application of ISO 55001

This chapter will undoubtedly be the most heavily sourced and technical chapter of this book. We will lay out the foundation of ISO 55000 and all supporting standard documentation to provide a feel for what is expected. Will you be an expert after reading this chapter? Probably not, but with the information added in the following chapters you will be able to form a very successful asset management plan around physical assets. Again, this particular chapter is just a short synopsis of the ISO 55000 series of standards. More detail will be unpacked in the following chapters.

If you are in a company that is heavily invested in Total Productive Maintenance (TPM), do not lose heart. Asset management and these series of standards do not supersede the TPM methodology. If anything, ISO 55000 enhances TPM and provides the catalyst you might need to drive forward with your strategy.

Let's get started.

ISO 55000

As mentioned, there are three standards that make up the ISO 55000 series of standards. The foundational standard is ISO 55000. Although this particular standard is understandably vague, it provides the necessary underlying context to start building your asset management program.

The purpose of putting such a standard together in the first place is explained in the opening text. Reading into the 'official' language of this international product, it is clear that the global community of standards entities felt it necessary to institutionalize the knowledge that the world possesses on such things. In its introductory segment, ISO 55000 speaks to the international cooperation between countries and the 'standards' organizations within those countries as the authority for compiling this list of mandated practices. ISO 55000 and the subsequent and compatible standards are very broad in scope. This is on purpose to provide some flexibility for the organizations that are adopting and seeking certification for asset management.

My faith in this approach was greatly enhanced when I noted that some of the sources for these books were standard reading for any reliability and maintenance professional. Source documents for ISO 55000, 55001, and 55002 include works from Ron Moore, Terry Wireman, and Ramesh Gulati. I specified this so that you, the reader, will appreciate that nothing mentioned and required in these standards should be unknown to us. It is literally the conversation we've been having for decades about how to best care for our equipment. Now the word is out to the masses.

Target Audience

If we fail in the implementation of asset management, I believe it will be because of this particular section right at the beginning of the standard. ISO 55000 lists the primary targets for the creation and deployment of such a far-reaching directive. Specifically:

- All those engaged in determining how to improve the returned value for their companies from their asset base (this means all assets, but we are focused on physical assets)
- All those who create, execute, maintain, and improve an asset management system
- All those who plan, design, implement, and review the activities involved with asset management

Why my concern? Read that bullet list again and ask yourself if that sounds like the responsibility of the maintenance department. If we are not careful, as an operating plant, a facility, or an organization, we are very likely to make asset management a maintenance program. Read further into the standards and you will understand why *everyone* needs to be engaged in managing the company assets. This is not a maintenance program!

Benefits of the Standards

To be certain, this is an international product. I mentioned at the start of this book that I was deeply concerned. I said that I was writing this book under duress. This section on the benefits was the root of my angst. I even sarcastically stated that all we needed to do to get the attention of upper management in regards to caring for the plant equipment was to write an international standard.

All kidding aside, there are tangible benefits for everyone from getting them on board and identifying their specific roles and responsibilities as they relate to developing a standard approach and training to conform to the stated practices. ISO 55000 states that the importance of an asset management system is that it is sustainable. In order to be sustainable, there has to be some assurances that the objectives of the organization are being met both effectively and efficiently. Isn't that what most of us have been pleading for over the decades? This is our chance to do great things together as a plant, a company, and an organization.

Influencing Factors

There are a number of influencing factors regarding the organization's operating environment, budget, and expectations from the stakeholders. All these are taken into consideration when we set up an asset management system and execute the required activities. We also monitor and take into account the environs of our particular situation by maintaining the system and being engaged in a process of continuous improvement to address risk and capture opportunities.

The governing idea behind the asset management standard is that the asset plan that is created by this structured and well-vetted approach essentially will translate the company's goals and objectives into decisions, plans, and executable actions. This must be accomplished using a risk-based approach.

That last line should be a dog whistle to each of us to adopt a risk-based approach to asset decision making. It has been my experience that very few organizations actually have such a technique being utilized in their facilities. Keep in mind that:

$$\text{Risk} = \text{Probability} \times \text{Consequence}$$

Risk is reduced by reducing probability (frequency) and/or consequence (severity). Figure 2-1 is an example of a risk matrix. Let this, or something similar, be a guiding element in all your decision making.

In clearer appreciation for this risk-based approach, ISO 55000 states that it is through the management of the risk, and the capture of opportunities that an organization can realize value in their assets and have a desired balance of cost-to-risk-to-performance.

SEVERITY \ FREQUENCY	FREQUENT (A) ≥1 per 1,000 Hours	PROBABLE (B) ≥1 per 10,000 Hours	OCCASIONAL (C) ≥1 per 100,000 Hours	REMOTE (D) ≥1 per 1,000,000 Hours	IMPROBABLE (E) <1 per 1,000,000 Hours
CATASTROPHIC (I) Death or permanent disability Significant environmental breach Damage >$1M, downtime >2 days Destruction of system/equipment	HIGH	HIGH	HIGH	MED	ACCEPT
CRITICAL (II) Personal injury Damage >$100K and <$1M Loss of availability > 24 hours and <7 days	HIGH	HIGH	MED	LOW	ACCEPT
MARGINAL (III) Damage >$10K and <$100K Loss of availability >4 hours and <24 hours	MED	MED	LOW	ACCEPT	ACCEPT
MINOR (IV) Damage <$10K Loss of availability < 4 hours	ACCEPT	ACCEPT	ACCEPT	ACCEPT	ACCEPT

Figure 2-1 A risk matrix

Adapted with permission by Marshall Institute, Inc.

The directive from this initial standard's overview is that top management, employees, and stakeholders are enlisted into the effective management of the assets. Again, we are speaking of physical assets, but the standard is applicable to all forms of assets, as the company identifies them. The individuals and groups just mentioned are responsible for establishing and executing control activities, monitoring those activities and working to minimize risks against the asset losing value, and again, capitalizing on those opportunities that would increase the value. Such control activities include, but are not limited to processes and policies.

Let's pause here and give this some deeper thought. "Top management" is required to implement planning control activities, monitor those activities, and reduce risks to an acceptable level. The standard purposefully keeps the definition of "top management" vague. It can mean the upper echelon of a plant staff or the C-suite executives of a corporation. The standard identifies top management as the highest person or people who has or have authority to direct and control the resources of the organization.

On the few lines that follow, list the ways your top management implements planning control activities to reduce risks:

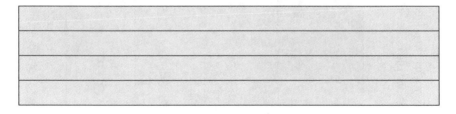

What comes to mind is an executive leadership body that insists on reliability upgrades and maintenance enhancing preventive and corrective maintenance. They (the top brass) effectively manifest this insistence by requiring that production stop for maintenance activities and participate in all forms of operator care, or autonomous maintenance. This is not often the case. For implementation of the ISO 55000 standards, the top leadership often needs to change their view of the priorities.

The standard lays this requirement on everyone, even the employees. Repeat this exercise again. In the space provided, list the ways that the production supervisor might implement planning control activities to reduce risks:

In this example, I'm reminded of the production supervisor who insists that maintenance PMs be performed on time. Also, this would be a supervisor who diligently requires that operators keep their equipment clean and run their machinery properly. This supervisor is someone who is *engaged* in the process.

Here is a final thought for this section. The standard's language quoted in an earlier paragraph mentioned that these actions are necessary to bring risks down to an "acceptable level." When a standard calls out such a subjective phrase, it is contingent upon the leadership to define (in writing) what the acceptable level is. In this case, there is a need for a documented interpretation for "acceptable level." We touched on this phrase in the previous chapter. At your location, what do you understand the phrase *acceptable level of risk* to mean? Record your thoughts here:

Fundamentals

To be effective, any standard, policy, or process has to have fundamental elements. Those elements must be present (meaning they exist, in writing), must be consistently deployed or executed, and must lead to a positive result. I have shared with many of my classes that: *average* x *average* x *average* does not equal world class. Neither, I say, does: *average* x *world class* x *average*. Fundamentals are foundational. A professional baseball player who cannot catch (a fundamental), will not be a very successful teammate nor have a long career in the majors.

The fundamentals as outlined in ISO 55000 include:

- Value
- Alignment
- Leadership
- Assurance

Further defining these fundamentals first requires an introduction to some key terms and phrases that are used throughout all three standards:

- Organizational objectives
- Life cycle management
- Decision-making process
- Stakeholders
- Risk-based
- Asset management system

Let's look deeper into those four fundamentals.

VALUE

Assets, physical or otherwise, are the source of income for the company. Any asset should provide value to the company and associated stakeholders. It is important to make this distinction early; the focus of asset management is not the asset, but rather the value of the asset to the organization. Figure 1-7 is reprinted here to drive this last point home.

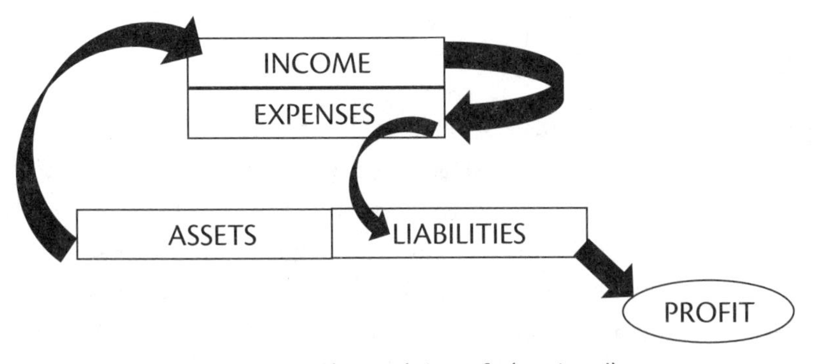

Figure 1-7 The result is profit (reprinted)

To make the asset-to-value connection, participants in ISO 55000 are required to present a clear statement aligning the organizational objectives with the asset management objectives. What are your organizational objectives? Have they been communicated to you? This will become a significantly important phrase. List your organizational objectives here:

Additionally, a life cycle management approach must be used to identify when and to what degree the assets will demonstrate the value that is anticipated. Remember, the life cycle of an asset doesn't end when it is no longer useful to your organization. The asset may have life and value at another facility. The term used to describe the asset's time of use in your facility is its *asset life*.

In terms of life cycle, or stages of its life cycle, ISO 55000 makes it clear that the organization must be conscious of the need for the asset, and the performance of the asset over the different levels (read that as 'evolutions') of its time in the organization. Furthermore, there has to be some analytical study of its performance and status during those stages of its life cycle. In other words, your organization has to have the metrics or Key Performance Indicators (KPIs) in place to determine if the asset performance is returning value.

Can you list three asset performance metrics used at your company that demonstrate value-to-performance? List those here:

1.
2.
3.

I can think of a few that might be helpful to consider:

- Cost of goods sold
- Conversion costs
- Maintenance as a percentage of total Replacement Asset Value (RAV)
- Total Effective Equipment Performance (TEEP)

The formula for TEEP is:

Utilization Time % x Availability % x
Performance Efficiency % x Quality Rate %

It is essentially Overall Equipment Effectiveness (OEE)
with the Utilization value added.

The description of the 'value' term concludes with a notice of the requirement to establish the decision-making criteria to be used. This decision-making process must reflect the stakeholder's needs and define the value of the decision being made. It will be interesting to note in the follow-on standards that 'stakeholder' refers to: customers, government regulatory agencies, and others who have a concern for the proper function of your organization's assets. This includes the employees of an organization. It is not overly common in the United States to ask the customer what they think of your asset management plan. Their input is not generally sought, but customers always seem to be the victim of whatever asset strategy the company comes up with.

The phrase or idea of a decision-making criteria is a new one to most folks; certainly it may be to your hourly employees. Has it ever been your experience that a decision is made at the top of an organization and those left to enact the decision are left to wonder why (or how) that decision was possibly made? In the few lines provided, explain how physical asset-related decisions are made at your location. What is the decision-making criteria?

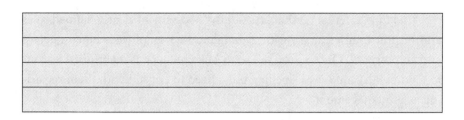

ALIGNMENT

If you've taken any of my classes, you have no doubt heard me say that the maintenance and reliability objectives, and now asset management objectives, should not be tangential, perpendicular, or parallel to the organizations' objectives. They have to be congruent and right on top of the organizational objectives. Anything different leads to competing priorities.

Alignment is an interesting concept and if you play out the thought, you'd most likely conclude that from time to time, things need to be brought back into 'alignment.' How that is done is the real question. The asset management standards suggest that any decision on how to align and how much to align is based on data (or information), risk assessments, and what is called a decision-making process or criteria. What is key here is that the alignment is made towards what the standards call the *organizational objectives.* This is perhaps the most referenced phrase in all three standards. Burn this idea into your brain because it's that important: *organizational objectives.*

Presumably this effort to align the asset management objectives with organizational objectives is supported by a locally designed and implemented supporting asset management system. The specification for such a support system should work to ensure that all the activities for asset management are in line with where the company is heading.

LEADERSHIP

It is not only *leadership* that is required to make asset management a part of the culture of the organization, but *commitment.* It is the commitment from the organizational leaders that is essential to improving asset management within the organization.

This fundamental section lays out the necessity of clearly defined roles and responsibilities along with a mandate that employees be competent and empowered. Once again, the term 'stakeholders' is mentioned in conjunction with employees, spelling out a need to consult with them regarding asset management.

Consider again the subjective terms *competent* and *empowered.* How exactly would you prove that your employees are competent and empowered? In the space provided, list several ways your organization ensures that its associates are competent and in what ways the company works to empower them:

ASSURANCE

ISO 55000 has an honorable intent, which is simply that the asset will do what it was purposed to do. If you will excuse the forced comparison, this definition is very close to Ramesh Gulati's textbook definition of reliability: "The probability that an asset will perform its intended functions for a specific period of time under stated conditions." (p. 53)

Asset Management vs. Asset Management System

Asset management has within its sphere the asset management system. The Venn diagram in Figure 2-2 shows this relationship.

Fortunately (yet vaguely), ISO 55000 gives us some insight into their definition of what asset management might be: generally, that all agencies of an organization will work to capture the value of an asset. Now, that's paraphrasing the standard, but in essence, the value that assets bring to a company is a result of all the work accomplished by all the people in the company. As a result, everyone ought to be pulling in the direction of ensuring that the asset will perform according to the plan, and that the value will be recognized, just as planned. This is a purposefully broad

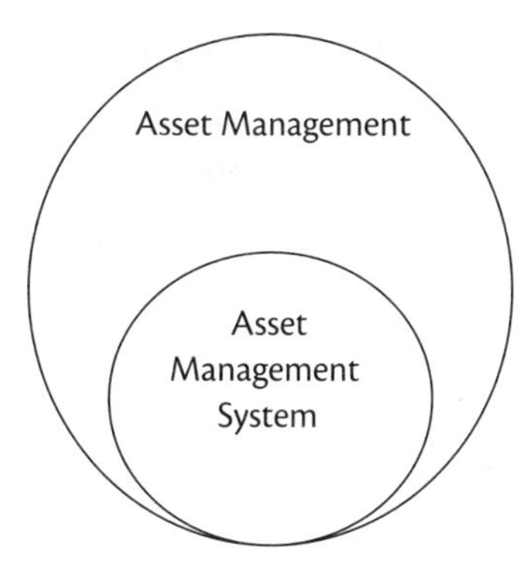

Figure 2-2 Relationship of asset management and asset management system

definition and gives companies an opportunity to establish the details for themselves. I believe it is helpful to think of asset management as "how do we intend to use this machine throughout its life *here* to get the value from it that we hope to get *here*." The word 'here' is italicized to make a connecting distinction to an early point regarding asset life and that an asset can provide value to another organization after your company is done with it.

We have an almost ready-made example of this very point that is alive and well with most people reading this book. If, for example, you purchased a new car and had the expressed intent on handing that car down to your son or daughter someday, how would you treat it differently than if you were just going to drive it until you wanted a newer car? Give that some serious thought. Would you take care of the car any differently? The key to the different approaches is the phrase *expressed intent*. If your company buys an asset with the expressed intent of getting some trade-in value, or resale value out of the asset later, would it be cared for any differently? I think it would. Manage your company's assets like your son or daughter will be driving them someday.

As seen in Figure 2-2, the asset management system is within the sphere of asset management. The asset management system is the result of policies and objectives centered on gaining value from the assets. It should be obvious by now that running equipment, or assets into the ground does not provide *value* to the company, nor can the asset provide future value if the asset is worn out and broken down. The asset management system is made up of dovetailing elements that exist within an organization. This is a subtle yet vitally critical point. A successful asset management company does not need to create a new asset management system. Rather, the company needs to take its existing systems and focus them on asset management. Your organization has an HR department, safety, accounting, procurement, etc. Are all of these agencies currently focused like a laser beam on asset management? That was a rhetorical question; I know the answer.

Consider for a moment the antithesis of an asset management approach that we have now, almost by default, for not being serious about this. Here is the example of our default settings (meant tongue in cheek):

- Asset management policy—run equipment until it can't run anymore
- Asset management objectives—have lousy PMs, don't stock the right parts, provide no down time for maintenance
- Processes—don't have any documented reliability processes

Now I ask you, who in their right mind would set up an asset management system that reflects those attributes? Warning: it might be you.

The asset management system is really a set of tools that is inclusive of plans, policies, processes, and information (or data gathering) systems that are all used to ensure that the asset management activities are being delivered. As you can imagine, with all applications of standards and regulations and mandates, documented evidence is sacred and absolutely required.

The approach laid out in ISO 55000 and the complementary standards provides a latticework structure to develop the system and support it, and to improve on it. There are many key benefits to this defined approach:

- Greater appreciation for data collection and interpretation
- New voices are heard on how to create value from the assets
- Cross-functional teams consider all aspects of an asset's life cycle
- Top management gains new perspectives and cross-functional teamwork
- Financial alignments are enhanced through data linking performance variables with their fiscal effect
- Human resource benefits through training and competency improvements
- Communication up and down the organizational chart improves

The many benefits of the asset management system are realized by the strategic execution of the many elements of the system.

The Elements

ISO 55000 spells out in broad terms eight elements that are key to building an asset management system. These elements have the ability to make a program successful or just a flash in the pan. They impact and require attention from the whole organization, including stakeholders, and demand that silos between divisions be relaxed so there is but one set of standards for the organization. The elements to engage in are:

1. Internal and external operating environment
2. Leadership
3. Contingency and opportunity planning
4. Resourcing and communication
5. Executing or operation
6. Auditing and determining performance to the process measures
7. Improvement
8. Working with and within other organizational systems

INTERNAL AND EXTERNAL OPERATING ENVIRONMENT

The section explaining what 'context' actually means is far reaching yet provides those adhering to this ISO standard enough flexibility to build an effective approach. Context, as we are instructed, is divided into external and internal.

External context includes, but is not limited to:

- Social
- Cultural
- Economic
- Physical environments
- Regulatory
- Financial

Where internal context includes, but is not limited to:

- Organizational culture
- Environment
- Mission
- Vision
- Organizational values

For the internal context of 'organizational culture,' what can you say about the culture at your location? Record your thoughts here:

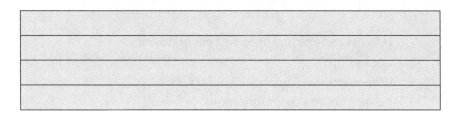

This element continues by addressing the importance of the stakeholder to the asset management system. The inclusion of stakeholders to the equation of determining how an asset is to be used to derive value to an organization will undoubtedly be an unusual addition. The asset management standards make it clear that the stakeholders are to be fully engaged in setting organizational objectives and asset decision making. This will be new and is understandably a huge scope of inclusion.

This inclusion from ISO 55000 really needs to marinate in your mind for a moment or two. When setting organizational objectives, the requirement is to involve the stakeholder, which includes (from earlier): customers, employees, government regulatory agencies, and others that have a concern for the proper function of your organization's assets. When was the last time your company invited the customer of your product or service into the boardroom to get their input on how the company assets should be managed? There are some progressive steps to be taken in adapting ISO 55000 to any organization. As mentioned many times before, when developing these nuances, documented evidence of application is a key requirement to satisfy any auditor.

LEADERSHIP

Top management is tasked with the heavy lifting in the ISO standard application. Keep in mind that this synopsis is on the asset management system. It is the absolute and necessary duty of the top management in any organization to define the values and the vision of the organization, and to set the roles and responsibilities, accountability, and the strategies. This cannot be completed in the typical top management double-speak; it has to be a message that is understandable to all.

Top management has an unambiguously assigned responsibility to develop the policies and objectives that will form and guide the asset management system. To be consistent with this, keep in mind that this is all accomplished to keep alignment with organizational objectives.

We first see the phrase 'organizational design' in this element. The top management must ensure that the organizational design (organizational flow and staffing) is capable of performing the duties and executing the elements of the overall asset management approach. No double-speak, just plain talk!

CONTINGENCY AND OPPORTUNITY PLANNING

A new term is introduced in this element, *Strategic Asset Management Plan* (SAMP). A SAMP may be recognized by other names such as asset management strategy. It is essentially how the asset will be used over the course of its life.

The SAMP is the guide to developing the asset management plans. This is the 'stuff' that we actually do to the assets to make sure they are available and capable of performing the work to create the value. Everything that is done to the assets, or equipment, must be measured for its effectiveness. I can't stress this fundamentally critical point enough. We have to measure results to determine if our actions are helping or hurting the asset's ability to 'make us money' and to meet other organizational objectives.

For the SAMP, picture in your mind a calendar laid out that has listed on it the dates of every single activity that a particular asset will be engaged in on specific dates for years to come. Imagine that calendar goes out for years, maybe three to five years. Now we are starting to think *strategically*. From the SAMP, the actual asset management plans are made.

What do you suppose would be in an asset management plan? Record some thoughts here:

These plans include the details of the activities and the performance measures and metrics to be used in gauging success, or concerns. These are measureable objectives and activities.

RESOURCING AND COMMUNICATION

Collaboration is key for this element. In fact, given that the pool of qualified candidates is shrinking and everyone is wearing multiple 'hats,' it is likely that sharing resources will become necessary, but will be a major sticking point for many organizations. Constant fine-tuning of the effectiveness and efficiency of human resource utilization is critical in this era of doing more often with less.

EXECUTING OR OPERATION

When an organization's asset management system is 'operating' and everyone and everything is going along with the plan, it is very probable that interruptions will still occur. It would be improbable to think that an asset plan could be effectively enforced without a hiccup or two in the span of the asset's twenty years of service.

The operation element advises the standards user that these changes to the plan provide an opportunity of risk 'introduction' into the formula. Some may refer to these episodes as 'periods' where you're running outside the curve and bad things happen.

It is even more necessary that the standards, policies, and directives be adhered to because of this possibility. This goes for outsourced labor and resources as well.

AUDITING AND DETERMINING PERFORMANCE TO THE PROCESS MEASURES

This may be a tripping point for many following the ISO standard. Few companies currently subscribe to an evaluation system that truly measures what is intended. We measure downtime when it is uptime we want. We measure stock outs when it is service level we want. Few, if any, maintenance metrics actually measure reliability. Instead, most maintenance metrics measure the effectiveness of maintenance administration. I bet you can't even name one reliability metric that your company tracks (that sounded horrible, but I'm serious—we're all to blame for this one). See if you can list any reliability metrics or measures that your company tracks. Record your thoughts here:

Now, consider the definition of reliability from Ramesh Gulati: "The probability that an asset will perform its intended functions for a specific period of time under stated conditions." (p. 53). Explain how the reliability metric you offered measures what is at the core of the definition for reliability:

I honestly hope that you proved me wrong.

ISO 55000 sets the requirement that asset performance will be measured, as will the adherence to the asset management system. These measures can be financial or not, and direct or not. The important aspect is to link asset performance to asset value realization.

The exercise concerning a list of reliability metrics was necessary to make this point. When we set out to establish performance evaluations, we have to be spot-on with our measures against what we are trying to do. Look at it through the eyes of someone conducting an assessment: "How does raising this metric improve asset reliability (or value)?"

IMPROVEMENT

Short and sweet. The asset management system is not meant to be static and never changing. In fact, constant movement towards improvement is necessary for organizational survival. Additionally, as the assets (physical, capital assets) move in and out of an organization, it would stand to reason that the overarching and guiding system for the control of the assets ebb and flow with the contents of the facility.

WORKING WITH AND WITHIN OTHER ORGANIZATIONAL SYSTEMS

This is a crafty element and pure common sense. Instead of creating another system to manage this new asset management effort, companies would be better off adopting, adapting, and integrating other management systems to control the asset management enterprise. This makes fiscal sense and would allow associates to work within the systems they are familiar with.

The drawback may be obvious. Several clients that I've worked with in the past have separate systems for maintenance, HR, accounting, finance, and production. Even if they are running an EAM (Enterprise Asset Management) system like SAP, it is likely that maintenance barely gets by on the maintenance module, but they (maintenance) have really gotten good at Excel!

We need to step up our game as far as systems knowledge and capabilities go in order to comply with the intent of this helpful element.

It was mentioned in the beginning of this chapter that there are actually three governing standards that make up the ISO 55000 family of documents. We just covered ISO 55000 Asset management—Overview, principles and terminology. Let's look deeper into the elements of an asset management system in a discussion of ISO 55001 Asset management—Management systems and requirements.

ISO 55001

When researching this standard, I had in mind an initial thought of the U.S. Constitution. I realize that sounds strange, but hear me out. ISO 55000 is an overview that lays the groundwork and general understanding. But, like the Constitution, it isn't complete. Ok, here is the Bill of Rights, and then continuous addendums through Amendments ratified by the states. ISO 55000 on its own is not complete. The addition of 55001, and 55002 later, work to give flexible, yet laser-like focus on what is required. I didn't like this standard at first. Remember, "I'm troubled. It's more like I'm concerned really," from the introductory chapter? However, I'm warming up to the idea. I'll tell you why.

In ISO 55001, the simple word *shall* is introduced. It is used quite often in this follow-on to ISO 55000. This standard document is the mating document to 55000 in that it adds thought, context, and direction to the asset management system's elements we were discussing in the previous section. This is a very good document.

The word 'shall' is first introduced in this middle standard, and so is the first mention of MOC, or Management of Change. There is also in-depth reference and explanation of the SAMP, or Strategic Asset Management Plan. These are all heady subjects and the standard is very explicit.

After the introductory sections, ISO 55001 launches right into the 'shall' directives and it doesn't let up for its entirety. As an example, ISO 55001 says that the organization *shall* determine what internal and external factors are important when calculating an ability to achieve the asset's objectives. The standard goes on to state that the SAMP *shall* be in alignment with the organizational objectives. There isn't a lot of wiggle room to get out from under a 'shall' directive.

As it relates to assets performing at your facility or location for years to come, error free, what are some internal and external factors that we feel would be important to put into the calculation? List those next:

Internal:
External:

When I think of the phrase 'shall determine what internal and external factors are important when calculating an ability to achieve the asset's objectives,' I can't help but picture in my mind the obsolete technology and other 'parts' we have hanging on our machinery and at what incredible risk we run in our factories by leveraging out our good luck. What is our plan to replace outdated and unsupported 'parts' on our assets? Would it make sense that with our intentions of running the assets into the future, for many years to come, that we have a handle on this tiny aspect of the asset management plan?

That's just page 1 of ISO 55001!

The good news is that we get to set the boundaries and applicability of the asset management system. We also get to determine the scope. In the scope development we need to consider the internal and external issues as previously discussed. Here's a reminder of the Venn diagram shown in Figure 2-2.

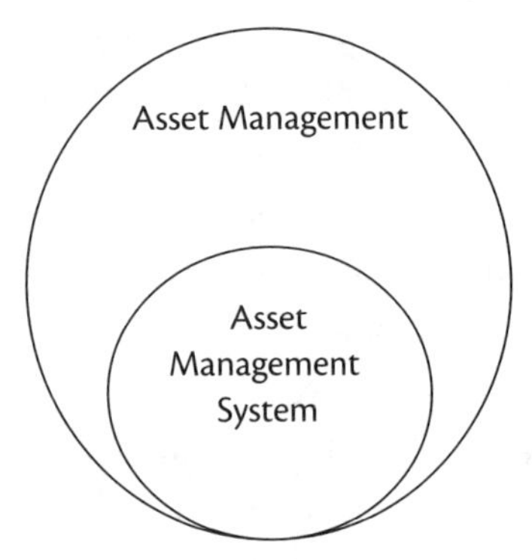

Figure 2-2 Relationship of asset management and asset management system (reprinted)

By further developing this diagram, we can gain a sense of these *boundaries* and *applicability* that we have control over. Figure 2-3 is the next evolution.

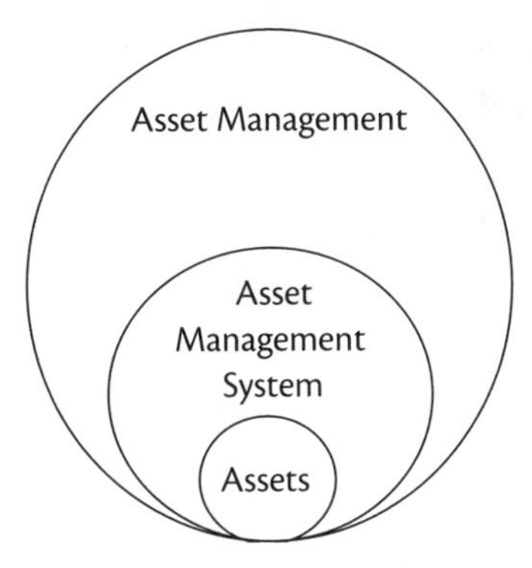

Figure 2-3 Assets within the asset management system

To recap this very important point: It is up to the organization to determine what assets make up the asset portfolio. Keep in mind the intent of ISO 55000 is for the organization to get value from the assets by way of asset management being in alignment with organizational objectives. Did I mention circular logic earlier? The needs and expectations of the stakeholders go a long way in determining the value to be gained and the route for the organization to move along to attain this value.

There is a lot of ground to cover in ISO 55001, but before we launch into that, I'd like for us to get a really good idea of what a Strategic Asset Management Plan is, as referenced earlier in this section. Darrin Wikoff says:

> A Strategic Asset Management Plan (SAMP) is nothing more than evidence of the "Documented Information" requirements and specifies how asset management objectives align, support, or will convert to strategic organizational objectives. In doing so, the SAMP also outlines the "Key Activities" required to achieve said objectives, which all asset management plans will subsequently include. Each activity is further defined within the plan relative to the resources required, cost considerations and timeline for implementation and realization of objectives. (p. 5)

In his message, Wikoff is specifically emphasizing two points: documented information and key activities. Duh! It's a plan, as in "what are you planning to do?" Now write that down. Use the game of chess as an analogy for the term 'strategic.' Always stay two or three moves ahead of your opponent. Don't only think of today's orders, but of next year's orders too, and what it will take to deliver them in the future.

I believe we've established a solid base of understanding from our glimpse into ISO 55001. I mentioned earlier that ISO 55001 provides a greater detailed explanation of the elements of the asset management system, and that this standard cites the word 'shall' very often. Let's start with the first element mentioned.

Leadership

Top management is directed through the phrase "shall demonstrate" to provide the top echelon level of leadership and commitment to the asset management system. How that manifests itself isn't clearly articulated.

The primary means of showing this leadership and commitment is to ensure that the asset management policy and SAMP are in line with the organizational objectives. By now you should recognize this as the guiding theme. An interesting note in this section also indicates that top management will integrate the asset management system requirements with the company's business processes. More than any other singular point in the standards, this last sentence squarely establishes maintenance and reliability as core value drivers of any business.

Wickoff warns us that, "Leadership and culture are determining factors when considering the organization's ability to administer the asset management system and achieve the organizational objectives, relative to value." (p. 28). I chose the verb 'warns' when I should have said 'forewarns,' but I honestly don't know the difference between the two terms. What I'm trying to convey is that our ability to achieve the organization's objectives as far as getting value from our assets is dependent on our organizational leadership and culture. You have been both warned and forewarned!

Record some thoughts to prove that we might have a long way to go on this particular risk. In the space provided, jot down some thoughts on how your top management and the culture of your company actually show respect and care for the upkeep of the production assets in your facility. Keep it clean, but seriously, freestyle some thoughts:

Leadership has to promote the asset management system, communicate the status, and encourage others to actively participate in the requirements and fundamentals. And as Wickoff points out, leadership must:

- Define roles and responsibilities
- Provide and allocate the resources
- Make sure stakeholders are informed, competent, and empowered
- Align the asset management system with stakeholder requirements

Contingencies and Opportunity Planning

ISO 55001 directs that organizations shall determine the risks that need to be addressed and the opportunities that need to be captured and manipulated. This thought is repeated several times throughout the three standards, yet no hard advice or direction is given. This is one of the positive aspects of the asset management standard. It is what we say it is. The intent is to prevent or minimize effects that are not desirable. What effects do you think might be undesirable? As it relates to assets, can you list one or two undesirable effects?

One of the most common undesirable effects is for the asset to fail. I had a young man in one of my classes tell me that their production equipment always seems to break in the middle of production. "Well, no kidding," I said, "I've never seen a machine that wasn't doing anything just keel over dead."

This is not only undesirable, but costly.

This section on planning includes some other interesting aspects of planning. Risks and opportunities can change over time. As such, the approach that the organization takes towards managing, reducing, and minimizing the risk, and gaining on the opportunities must change as well.

We don't usually give much thought to the positive opportunities that might come to pass within asset management. It seems like we're always just trying to stay one step ahead of catastrophe. My first civilian boss told me to always have a plan in case things go right. I had a plan all right, but I never got to use it. For me, contingency plans were usually retreat plans.

Resourcing and Communication

I love this section. A clear mandate is established that insists the organization shall determine the resource needs in order to establish, execute, and maintain the asset management system. What makes this section particularly helpful is that it continues to spell out that the organization will provide for meeting the asset management objectives and activities. I read this to mean 'the maintenance of the assets.'

This section dovetails nicely into points made earlier. Not only are we required to have the proper resources to achieve the organizational objectives relevant to asset value, but those resources have to be (and you've seen these before):

- Competent
- Aware
- Communicated with
- Informed

'Informed' actually means information requirements. The organization has to (shall) determine what information techniques and systems are necessary to communicate to the masses on what is required for asset management, how the organization is doing, the achievements or challenges being faced and actions taken.

Recall from ISO 55000 that to have a fiscally responsible, efficient, and effective asset management system, the asset management system has to integrated into the organization's other systems. Also recall that *average x average x average* does not equal world class. We have to step up our game in terms of our CMMS (Computerized Maintenance Management System) and our overall information and communication processes. We have to gather and understand what the machines are telling us.

Wickoff, in his work, recognized that organizations must demonstrate their understanding of the very data they collect to make educated decisions affecting asset management objectives and plans. This is the exact interpretation of what the standards call *data-based.*

Executing or Operation

This is the section that refers to MOC (Management of Change) by requiring that risks associated with any planned change might have an impact on achieving the asset management objectives and shall be assessed before the change is implemented.

As mentioned in *The Reliability Excellence Workbook: From Ideas to Action*, "A MOC actually provides two functions. Primarily, the MOC is a coordinated, formally documented process that is used to determine if a 'change' is technically feasible." (p. 171). And, "A MOC is also used to inform all the concerned parties that a change is taking place, so it becomes a means to communicate." (p. 172).

This element on 'operation' also includes the requirement to implement control of the processes and to keep documented information to provide evidence that the processes have been carried out properly.

Auditing and Determining Performance to the Process Measures

I made a very bold differentiation in my first book and it was on the non-similarity of metrics and KPIs (Key Performance Indicators). I'd like to share that with you here because I believe it is an important point of distinction.

"A metric is often a measure of something we take at a set sampling period to determine its magnitude. This is typically a number (or value) that we track to gain a sense of the trend of an event or activity. A KPI, or Key Performance Indicator, is a type of metric that is tied directly to a stated (and therefore, written down) goal or objective." (Ross, p. 84). Figure 2-4 shows this relationship.

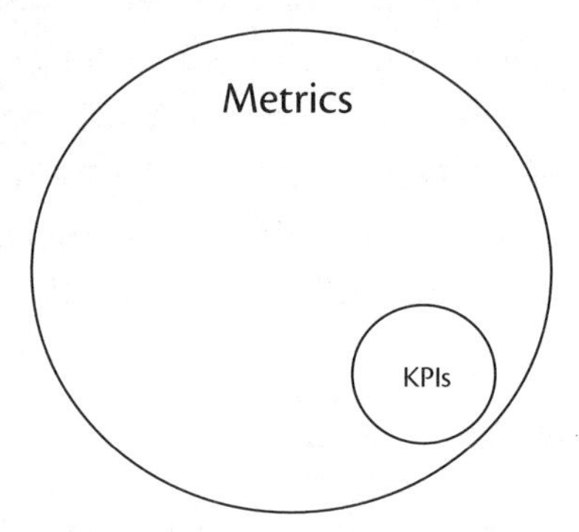

Figure 2-4 A KPI is a type of metric

The point I had hoped to drive home here is the fact that our SAMPs (Strategic Asset Management Plans) are our documented intent for extracting value from our asset's performance. This is a documented plan. A KPI is a metric tied directly to a stated (read that as 'documented') goal or objective. Recall that organizational objectives are the theme of the ISO 55000 series.

To stay on track with the asset management plan, the system has to be continually audited and assessed through management reviews. Audits and other such inward-looking exams are meant to keep the processes moving in the right direction and to control any straying from the intended path. Corrections are often needed and quite necessarily expected.

Improvement

This is a very short and to the point element and closes out our discussion on ISO 55001. Without exception, my favorite part of ISO 55001 (and I do realize how silly that sounds) is the passage that indicates that when an incident occurs (breakdown) on one of the assets or with the asset management system itself, the organization *shall* take action to correct it and to deal with the consequences. I'm not sure why that thrills me to see that in print. If we're not careful in our facilities, though, this might just turn into a lot of pencil-whipped 5-Whys.

As ISO 55001 adds detail and greater depth to the foundation laid in ISO 55000, so does ISO 55002 build upon ISO 55001. The complementary standards are the nucleus of asset management.

ISO 55002

We're going to deviate slightly from the course taken during our discussion on 55000 and 55001, and dive a little deeper into context for our study of ISO 55002. Primarily this is needed as this third and last standard really begins to form what your asset management process might look like.

Interestingly, the word 'shall' is relaxed in this third offering and is replaced with the more subtle, 'should.' This section won't be as long, but it is important that the reader (of both this book and the standards) gains a sense of what is included in ISO 55002.

Overview

The organization, your organization, has a management system. This could be a really good system or one that requires a lot of attention. Regardless, the asset management system should be an integrated part of the existing system. This merging of systems needs to be in accordance with and in line with organizational objectives and the organizational plan.

ISO 55002 continues by listing what the asset management system includes. It implies that asset management must have:

- A policy
- Objectives
- A SAMP
- Plans

I mentioned in the introductory chapter that I wondered if we're even capable of doing what we're being asked to do. My concern was that we haven't managed to do it yet. The basis for this concern was a passage from ISO 55002 stating that the organizational objectives are born from the organization's strategic plan and the documented activities within that plan. My first book was written to address the fact that I don't believe we have a strategy at all. The application of the ISO 55000 standards are predicated on the fact that we (that is, our organizations) have some strategic plan. If it exists at all, it is a sure bet that the strategic plan is held close to the vest at corporate headquarters. This is a great concern.

The asset management policy should lay out the principles that the organization intends to employ through its asset management process in order to achieve the organization's objectives. The approach to enacting those principles is documented in the SAMP.

A sidebar is necessary to address this issue of the 'organization's objectives.' This phrase resonates throughout the three standards. Do you have any idea what your organization's objectives are? Are they written down and communicated? Not your exact plant or facility's objectives, but the *entire organization*. List any organizational objectives you have knowledge of here (note you made reference to these earlier in this chapter):

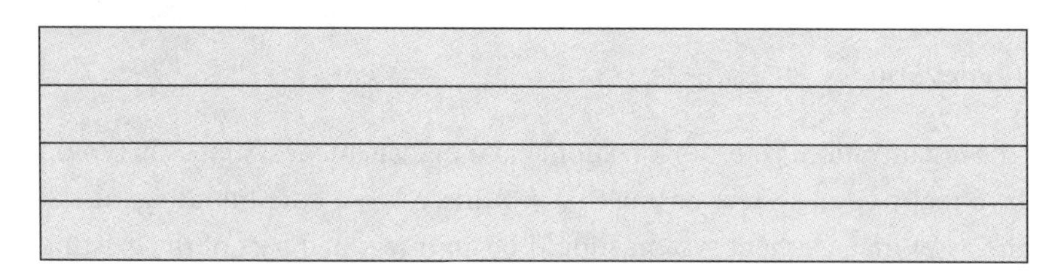

If you aren't aware of these objectives, that would need to be the first hurdle to tackle.

The SAMP mentioned a few paragraphs before is the documented link between those organizational objectives (did you record any?) and the asset management objectives. The SAMP defines the framework needed to accomplish the asset management objectives. Not to be confusing, but there is an asset management plan and then there is SAMP. The difference? SAMP is strategic, and looks years into the future and focuses on alignment and therefore 'positioning.'

The SAMP (think strategic) is a very high level view of the asset and how it contributes for years to come to the bottom line of the organization. This SAMP lays out all the known activities that an asset will experience for the foreseeable future. These asset level activities are the actual actions your company takes to preserve the asset so it can add value to the company. This is the stuff that we do.

ISO 55002 lists some of the developmental thoughts that should go into the SAMP:

- Expectations of stakeholders
- Activities that extend beyond the organization's planning time frame
- Decision-making criteria

Do you recall the story I shared in Chapter 1 about my friend in the Air Force motor pool and how the Air Force determined their maximum spend limit on each car? That is clearly documented, asset-related decision making.

Our asset management system has to be aligned with what the three standards refer to as internal and external context. This really means that we have to be aware of forces and interests inside and outside of our organization. These have been listed previously.

External context includes, but is not limited to:

- Cultural, political, legal, regulatory
- Values of external stakeholders

The internal context includes, but is not limited to:

- Oversight and governance
- Organizational structure
- Policies

In this listing and throughout the discussion of these three standards, I want to drive home the point that when you see the term 'stakeholder,' the definition is 'almost everyone affected by our decisions.' It is a long list, including but not limited to:

- Associates (including union leaders)
- Other groups and divisions within the organization (e.g., engineering, accounting)
- Shareholders (owners, stockholders)
- Customers
- Suppliers
- Local communities
- Societies interested in environmental sustainability
- Insurers

I believe there are exactly two issues that will kill any attempt at ISO 55000 implementation. They are (in order): 1) Making it a maintenance program, and 2) Failing to engage (fully) all the stakeholders. Keep in mind that you need to document the stakeholder's expectations.

If you're still reading this book, you have to be exclaiming, "Holy cow, this looks like a lot of work!" And it is. The fortunate aspect is that *you* determine the assets to cover and the scope of what is included in your ISO 55000 program. These are boundaries that have to be well defined.

As it turns out, the organization determines exactly what is and what is not included in the asset management system. To be certain, as it relates to physical assets or capital assets, a *scope* can be introduced in the SAMP addressing these boundaries. These boundaries are to be communicated to the internal and external stakeholders. This is not a license to get off easy. For an asset management certification to mean anything, it has to have some 'heft' of importance to the organization and to the stakeholders. Do the right thing here and cover what is important to your customer.

This revelation is actually a saving note and brings a level of (pardon the term) sanity to a very complex and administratively overwhelming process. We—that is, you—get to decide what part of your organization to fold into an ISO 55000 certification.

What assets would you believe need to be folded into the formal asset management program? Provide a short list here as an example:

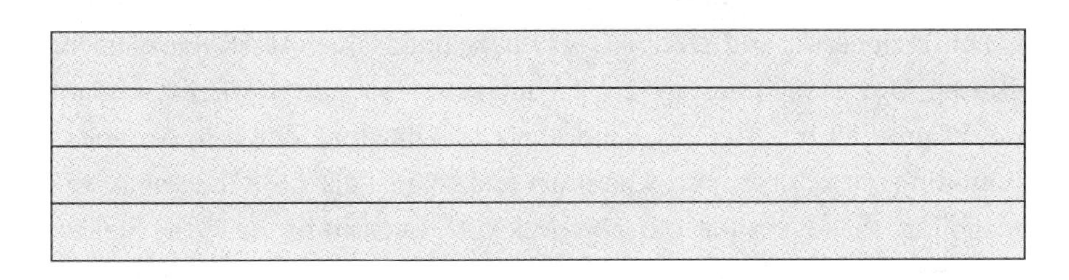

I want to conclude this overview section with just a thought about professional associations. It's a sure bet that your company, at some level, belongs to a professional group. If you're in the food business, the pharmaceutical business, steel, automotive, etc., your company should belong to an associated fraternity. It is important to have these relationships and opportunities to network and keep current on what the customer wants and the industry demands. Get out and seek out those relationships.

There should be little doubt that this ISO 55000 effort needs to be driven from the top. This requires a lot of managerial skills, but an ISO 55000 certification is not obtained by managing your way to success; the effort must be led.

Leadership

Do you recall what I said were the two surefire ways to kill an ISO 55000 effort? Number one was to make it a maintenance program. Here we go!

ISO 55002 states that top leadership are the influencers of the asset management process. It is these top leaders who can appoint an individual to have the responsibility for the system and the entire enterprise of asset management. However, there is a stipulation that overall accountability remain at the top.

This should smell like delegation to you. Do not make this a maintenance program.

I would submit that most organizations are under pressure to keep indirect or overhead costs down. In fact, I know of few organizations that are trying to build their salaried ranks above what they are budgeted for. There are undoubtedly holes to fill in most organizations, but having more weight at the top of the organizational chart is rarely desirable.

A colleague once told me that companies that are serious about asset management will have an asset manager in the C-suite executive level. Not a

Chief Engineering and Technology Officer, but a Chief Asset Management Officer. Our companies are capital intensive; no one is stirring a giant pot of product out there by hand, there is a machine doing it. No one is pounding out a horseshoe by hammer and anvil; a machine is doing it. The machines, the assets, pay our paychecks. We need someone at the highest level making sure that asset care is a priority all the way from the shop floor to the boardroom and back.

Why the dissertation on this subject? Buried in ISO 55002 is a little paragraph that states that it is often necessary, when confronting real budgets, to have lofty plans but have to defer some of them for another time. The standards speak of the process of trying to zero in on fulfilling all the calculated needs as an *iterative* process. This would be a constant reconciling of the needs against what can be afforded at the time. The standard goes on to 'suggest' that maybe, in light of the restricted human resources that are available, or allowable, perhaps the organization would relax what activities are proposed. I liken this to telling your boss you don't have enough maintenance people to do all the PMs and expecting your boss to be 'ok' with that. The trouble with *iterative* processes is they have a way of losing momentum as attention is drawn elsewhere.

This was a fancy paragraph to state the obvious: "You can't always get want you want." (Jagger, Richards, 1969).

The good news is that top management is clearly tasked with making ISO 55000 work in the organization and in line with the organizational objectives. It might take a compelling argument, but eventually the top brass can be made to understand that asset management is different than maintenance and reliability. Maintenance and reliability are parts of the asset management plan (remember the 'activities, the stuff we do').

Leadership has to be committed to clearly defining which role is responsible for which of the activities. The preferred method is through the development of job descriptions. This doesn't have to be a complete re-write of the great job descriptions you may already have (said tongue in cheek). Where necessary, just add the asset management responsibilities to the list already created and fielded.

The detailing of roles and responsibilities and holding people accountable for what they are entrusted to do should keep the organization moving along the right path. However, sometimes best laid plans are for naught when issues arise. Through effective leadership, organizations are better able to handle contingencies and 'right the ship.'

Contingency and Opportunity Planning

Most organizations have a risk management approach, albeit a very poor one. I would suspect at the upper levels of an organization there are some very detailed steps to avoid or minimize risks and it is likely to be almost 100% financial exposure or brand protection. These two aspects can sink a company quickly—even a big company (Enron?). However, how much risk avoidance is part of the daily makeup of an organization when it comes to asset management? I contend that we do the opposite on a daily basis by making the situation worse, and increasing our risk. Recall that: Risk = Probability $\times$ Consequences.

As ours is a risk-avoidance profession, it should not come as a surprise that the standard actually requires organizations to take action when a risk is discovered or calculated. This is not only relevant to the assets themselves, but the asset management system as well. This section brings back for discussion the adaptation of a risk matrix; an example was provided in Figure 2-1.

Organizations are further tasked in ISO 55002 with evaluating the effectiveness of the actions they do take to eliminate, control, or minimize risks. The term used is to 'manage risks' and to demonstrate that you have done so. Figure 2-5 is an example of such a demonstration. This graph

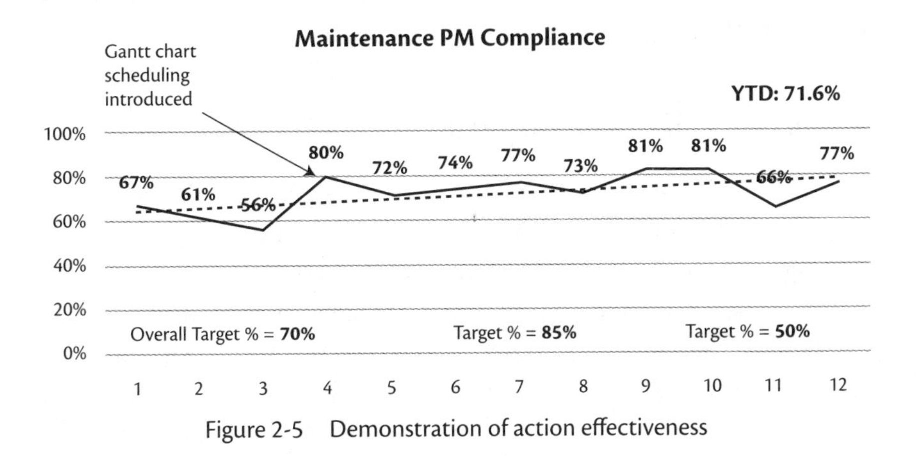

Figure 2-5 Demonstration of action effectiveness

shows an improvement of a less than high-performing PM compliance effort. The spike was the introduction of a Gantt-charting operation conducted by the data analyst. Note the absolute objective evidence that the company took action, and the almost immediate (positive) result of that action.

Of course, we don't just plan for contingencies, we also plan to make sure things are going to go right in the first place.

ISO 55002 provides an interesting twist to contingency and opportunity planning that I suspect not many organizations have experience with, and it centers on documentation and communication. Specifically, it may be typical for asset plans to explain the purpose for an approach being enforced, plans that are laid out, and for the money being spent on maintenance and asset modification. This is often done against a measure of how successful previous plans and activities have been.

Remember that communication of "what we're doing and how we did" are key points to be kept out in front. This key inscription will be used as the basis for the discussions on production and maintenance's role in ISO 55000. This makes sense if you consider that all stakeholders are to be engaged and made part of asset-related decisions. This really is an all-inclusive communal approach to business.

All these plans need support for execution. There was an earlier mention of the 'resource' note to ISO 55002, in that we don't always get what we want. Truthfully, everyone has to operate within a budget. My major point was that asset management should not be a default responsibility of the maintenance department. However, 'support' means much more than just having the bodies.

Resourcing and Communication

You can't just have the bodies. New employees are a liability until they know how to do something. If ever there was a clear case for a rock-solid training program, it is in this element of ISO 55002.

Those associates have to have a level of competency in their roles. The standard isn't just addressing the competency of the maintenance and production hourly folks, but the leadership as well. Whatever the individual role is in an organization, those people must be capable of executing their piece of the asset management plan. For example, a tradesperson needs to be competent in asset equipment condition (through preventive maintenance) and reporting that condition to those making 'data-based' decisions. Give some real thought to how your maintenance associates prove their competency in executing PMs and communicating that information (with facts, figures, measures, and determinants) in a way that allows for 'data-based decisions.' Record here your thoughts on how your maintenance associates prove their PM competency and their skills at communication:

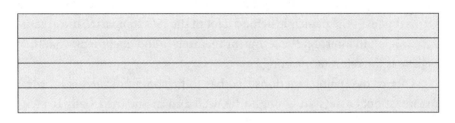

Competency is certainly a critical aspect to asset care and asset management. Another equally important element is that of general awareness. If you've ever spent time in one of my classes, or if I've consulted at your location, you have absolutely heard me talk about finding the "what's in it for me" argument in all the changes and improvement efforts that our organization undertakes. It is clearly the responsibility of the associates to make up their own minds as to what the value in any change holds for them, personally. But, as leaders, if we can't 'prime the pump' of thought in this direction, we should not expect change to be sustainable.

People have to be aware of what is expected in terms that they can understand and this will at least germinate an idea of some profitability for each individual. It was very thoughtful and forward thinking of the authors of the ISO standards, specifically 55002, to include a passage that requires leadership to make the stakeholders (including associates) aware of the asset management system and its requirements. How can someone support the ISO effort if they don't know what it's about?

This effort to make people aware of and find value in how they individually can profit from the adaptation of ISO 55000 is made possible through an organization's communication process.

Communication is the key to success in almost any venture. No, strike that, in *all* ventures. The problem, as I see it, is that we don't have a lack of communication in our organizations, we have lousy communication. There are some very soft and subjective terms in the requirements from ISO 55002 in terms of communication. The asset management activities, we are told, "should be periodically communicated to the stakeholders." This communication should be in conjunction with other elements of the asset management system.

There are a lot of mandates in the ISO 55000 series of standards that are subjective. The section on communication is just another example. What exactly does 'periodically' and 'should be' mean? These are subjective terms, but as is the principle behind a lot of the ISO standards, the organization is left to interpret the terms for themselves and document what they believe is the relevant meaning.

One of the major functions that has to be communicated is how the organizational assets are going to be used and maintained. This is mentioned in the 'operations' element.

Executing or Operation

Highlight this section of this book. This is the cornerstone of ISO 55000 that requires everyone to work together in the interest of having the asset performing at a high level to derive the highest value for the organization. ISO 55002 mandates that organizations conceive and execute planning and process control procedures to ensure that asset supporting activities are accomplished. Not only is that the mandate, but the qualifier "*effectively* accomplish" is required. The direction continues by including a process to determine who is to effectively accomplish each specific asset value-guaranteeing activity and to demonstrate how the actions help to reduce or minimize the risk. This is an exceptionally powerful passage. Perhaps the most powerful one in all of the three standards. The totality of what actions are necessary to keep an asset generating value in and for the organization is captured in the asset management plan.

Focus in on the 'asset management plan' and you'll quickly conclude that we have to have one. Where is your comprehensive, bare bones, 'we must do this, and do it this way,' asset management plan? Chances are you don't have one. I didn't either. But we need one.

When an asset is in its *operational life phase*, what methods are used to evaluate the condition of the asset and also, how do we measure the effectiveness of the evaluation process?

Auditing and Determining Performance to the Process Measures

I heard a long time ago that if you don't measure something you can't improve it. I say if you don't measure something you can't *prove* it. In our profession, we have a lot of faults, and we know most of them and are striving to improve. One area that I wish we had a magic pill for is this area of performance measure. In maintenance and reliability, we don't do a very good job of proving our worth to the organization. Someone in our organization looks at our work as an 'expense,' and not an asset. We can change that and ISO 55000 will help.

Many years ago, when I was the plant engineer of a Midwest manufacturing facility, I contracted the services of a predictive maintenance company. The thousands of dollars we spent each year with this company paid off in the simple fact that none of the eighty assets they monitored ever failed. We had a seven-year success story. A new controller was hired by my plant manager. One of the first things she wanted to cut to improve our bottom line was the predictive maintenance contractor. Her rationale? We didn't need the contractor because we hadn't had any failures in over seven years. We, in reliability and maintenance, need to do a better job calculating our worth and communicating it to the enterprise.

There are enough volumes and reams of information on metrics, measures, and KPIs to fill several libraries, so we won't discuss these elements. Rather, what might be lacking at our location are true performance measures of equipment condition.

The asset management industry has gotten beyond mere preventive and predictive maintenance. The next generation of our thought appears to be going the way of the IIoT (Industrial Internet of Things), AI (Artificial Intelligence) and Augmented Reality. Do you remember when barcodes were a big deal?

This real-time monitoring is complex and costly. A lot of the adaptations necessary to employ such technology are not compatible with much of the aged industrial equipment still in use. One thing is for certain. It would be foolish to collect this data, interpret it (through software) and then ignore it. How many times have you personally told someone that an asset was going to fail if you weren't allowed to stop and fix it? Would that person act more directly on the advice of a computer? The scary thing is that I think he or she would.

Despite our best efforts, things go badly sometimes. Rounding out this discussion on ISO 55002 and on the ISO standards is the element of 'improvement.'

Improvement

Remember that we had several discussions on how the asset management system needs to tie into your already existing systems and be congruent with them. This is another case where that would be economical and beneficial. For this element, give some thought to your RCA (Root Cause Analysis) process. The best example you might have at your location could be with the safety department.

ISO 55002 requires that organizations have the ability to recognize issues and concerns when assets or the asset management system aren't delivering the value intended, and to take immediate action. In fact, plans and processes must be pronounced to control these 'nonconformities' and to minimize the probability of recurrence or consequence.

Sometimes improvement actions are taken to maintain the gains and capitalize on the momentum. True, we have to assess and correct where things went wrong, but how about the opportunities when things go *right*?

Opportunities are the counterbalance to issues and concerns, or what the standards refer to as 'nonconformities.' As an organization has the responsibility to monitor and stamp out faults and errors, so too does it have the requirement to seek out opportunities for improvement and advancement. All this is made possible through routine and repeated monitoring for the potential to make improvements. 'Improvement' isn't just something that an organization does once a year, it is continuous. Hey, I just realized that it's 'continuous improvement!'

I'm concluding Chapter 2 with a helpful overview of the major portions of an asset management system. Remember, all these aspects have vast room for interpretation.

Documentation, communication, and constant auditing (performance evaluation) are the keys to success. More to come on that in the following chapters. Figure 2-6 is a relationship pictorial of what we have discussed in these three standards.

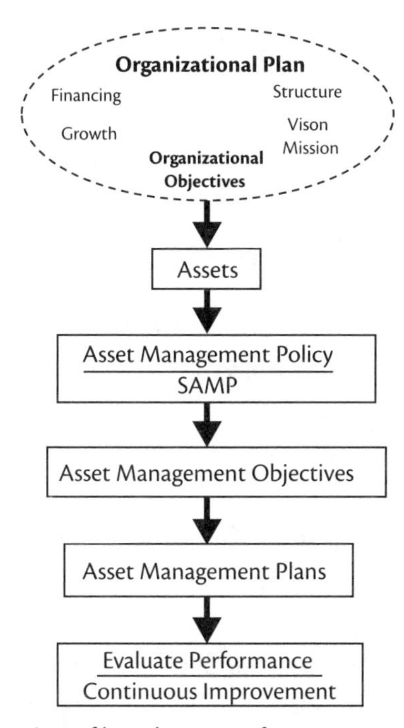

Figure 2-6 Overview of key elements of an asset management system

CHAPTER SUMMARY

Even I will concede that this was a long and dry chapter. It has to be evident to even a person with pedestrian knowledge of how the world works that in order to do anything, you have to first understand how it's done. From that you can modify it and make it your own, but to first understand the foundation is critical. Even a chef needs to learn how to follow a recipe first.

In Chapter 2 we discussed ISO 55000, 55001 and 55002. The foundation for our understanding of the asset management processes was developed in

55000, added to with the mandate of 'shall' in 55001, and 55002 rounded out how and suggested ways the elements could be executed.

There were some key words and phrases that are going to form the basis for our discussion going forward. If we cannot put our fingers on this information, chances are we will struggle and fail with asset management. Let's continue to build our strategy.

Adding to Our Business Case

For asset management we choose to adopt the suggested and often mandated principles set forth in the ISO 55000 series of documents: ISO 55000, ISO 55001, and ISO 55002. As such, our intended audience resides in one or more of the following areas:

- All those engaged in determining how to improve the returned value for the company from their asset base (we are primarily focused on physical assets)
- All those who create, execute, maintain, and improve an asset management system
- All those who plan, design, implement and review the activities involved with asset management

The primary benefit to adopting such a methodology is to ensure a sustainable approach to meeting organizational objectives with the performance of the organization's assets. We intend to adopt a risk-based and decision-making criteria. What follows is an example risk-based criteria; this is a risk matrix.

It is through the management of the risk, and the capture of opportunities that our organization can realize value in their assets and have a desired balance of cost-to-risk-to-performance.

The primary task of the asset management approach is to ensure organizational assets work to achieve the organizational objectives as applicable. Our understanding of the organizational objectives are:

SEVERITY / FREQUENCY	FREQUENT (A) ≥1 per 1,000 Hours	PROBABLE (B) ≥1 per 10,000 Hours	OCCASIONAL (C) ≥1 per 100,000 Hours	REMOTE (D) ≥1 per 1,000,000 Hours	IMPROBABLE (E) <1 per 1,000,000 Hours
CATASTROPHIC (I) Death or permanent disability Significant environmental breach Damage >$1M, downtime >2 days Destruction of system/equipment	HIGH	HIGH	HIGH	MED	ACCEPT
CRITICAL (II) Personal injury Damage >$100K and <$1M Loss of availability > 24 hours and <7 days	HIGH	HIGH	MED	LOW	ACCEPT
MARGINAL (III) Damage >$10K and <$100K Loss of availability >4 hours and <24 hours	MED	MED	LOW	ACCEPT	ACCEPT
MINOR (IV) Damage <$10K Loss of availability < 4 hours	ACCEPT	ACCEPT	ACCEPT	ACCEPT	ACCEPT

We will continue to show that the assets are performing in alignment with the organizational objectives by including the following metrics in our report on value-to-performance:

In addition to becoming an organization that reaches risk-based conclusions, we will also be an organization that is data-driven and follows the decision-making criteria listed here:

Our associates will be competent and empowered as evidenced by our practice of:

There are external and internal forces that make up the operating context of our facility. Those may include, but are not limited to:

External context:

- Social
- Cultural
- Economic
- Physical environments
- Regulatory
- Financial

Where internal context:

- Organizational culture
- Environment
- Mission
- Vision
- Organizational values

Our approach to establishing an asset management culture will be in line with the following Venn diagram:

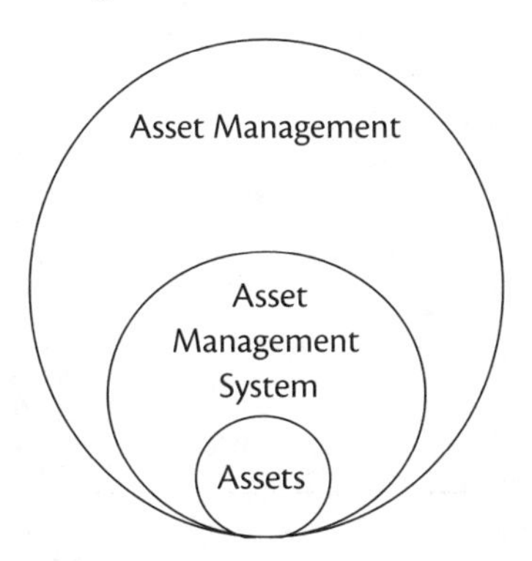

The organization determines exactly what is and what is not included in the asset management system. To be certain, as it relates to physical assets or capital assets, a *scope* can be introduced in the SAMP (Strategic Asset Management Plan) to address these boundaries.

These boundaries are to be communicated to the internal and external stakeholders. For our purposes, we have determined that the following assets will be under the auspices of the formal program:

Our asset management activities will contain contingency planning for when events occur, and opportunity-seeking efforts when those instances arise.

We are required to have the proper resources to achieve the organizational objectives relevant to asset value, and those resources have to be:

- Competent
- Aware
- Communicated with
- Informed

It is fiscally responsible and good common sense to build an asset management system from the systems that are already present in the organization. Those systems are:

What follows is a representation of the major elements to our asset management system:

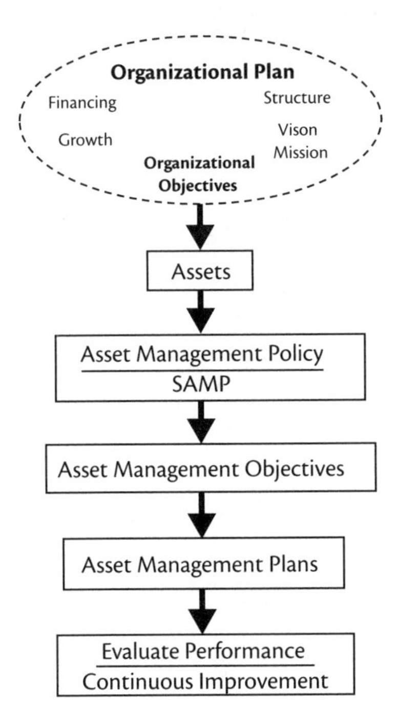

Turn to the last section of the book and record this same information in the comprehensive strategy section.

THREE

Capital Assets

"More than a convertible, the car had to be equipped with a retractable hardtop. I did not want to run the risk of the tops being slashed, which could put a damper on a trip around the world. Furthermore, I decided it had to have a diesel engine. Trucks, buses, trains and boats around the world run on diesel fuel, and you can always get it."

—Jim Rogers, *Venture Capitalist* (p. 7)

On January 1, 1999, financier Jim Rogers and his girlfriend Paige Parker set off on an around-the-world driving expedition from Reykjavik, Iceland. In three years, they would cover 152,000 miles, six continents, and 116 countries. The quote headlining this chapter is Jim's dead reckoning understanding that he needed a machine that would 'do what it's supposed to do, for as long as it was supposed to do it.' He was very particular about the assets he and his team would rely on for this incredible journey. Risk = Probability × Consequence.

He chose a modified Mercedes SLK roadster with a diesel engine and the chassis of a G-Class sports utility. His belief was that every third-world dictator has a Mercedes, so there must be a capability in every corner of the world to fix this make of car. On a journey of this scale, you must have the machinery that is manufactured and maintained to reduce risk. In 152,000 miles there is a high probability that something will go wrong, and the consequences in a less-than-hospitable country could be fatal. It all started with the right machine.

For this chapter on capital assets, we need to first determine at what point we should get a new asset or continue to modify the existing asset to gain more life from it. Buying new or modifying an existing asset are usually the only two options we have at 'getting' the right machine.

Like Jim, for us, we need to choose the right machine.

Buy New or Make Do

We've touched on this idea twice in this book. I mentioned my service friend at Wurtsmith, AFB in Michigan and his explanation that the Air Force keeps a staff car as long as it (the car) still has credit on its repair account. I also shared the story of my son asking when you know it's time to buy a new car rather than keep fixing the old car. The answer, my answer, was of course, "When *you* can afford it."

I'm going to bring Figure 1-8 back in for discussion (shown on the next page).

On this reprint of Figure 1-8, take a pen or highlighter and make a mark on the 'operating life cycle' arch of where you would decide to either modify or buy another replacement asset. No rules or advice to guide your marking, just make a little 'tick' mark on the figure.

In as few or many words as you need, record here why you made that mark there:

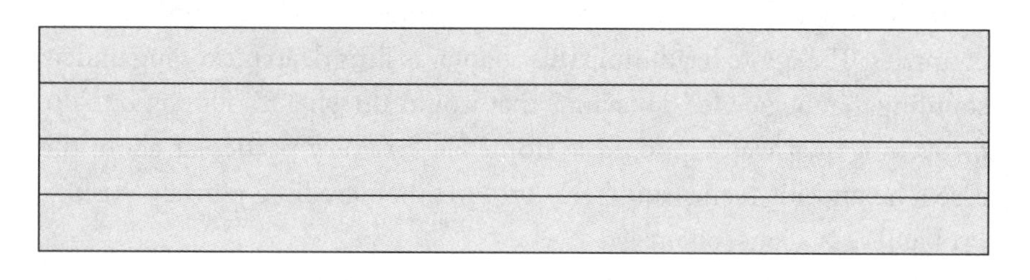

Determining if you should modify or replace an asset can be just as simple as the method you just used, or as complex as an algorithm running integrated cost-benefit ratios and further analysis. I'm curious what mix of methods we have running throughout our boardrooms and plant conference rooms today. This is part of the decision-making criteria we learned about in Chapter 2.

The answer to this puzzle, in a new-world order that is guided by asset management and specifically the ISO 55000 series of standards, lies in determining when the asset's value to the organization begins to show diminishing returns. Have you owned a car or house that you just kept throwing money at, or into? We have a term for that—"nickel and diming you to death." In business terms, our nickels and dimes are hundreds of thousands of dollars, if not millions of dollars.

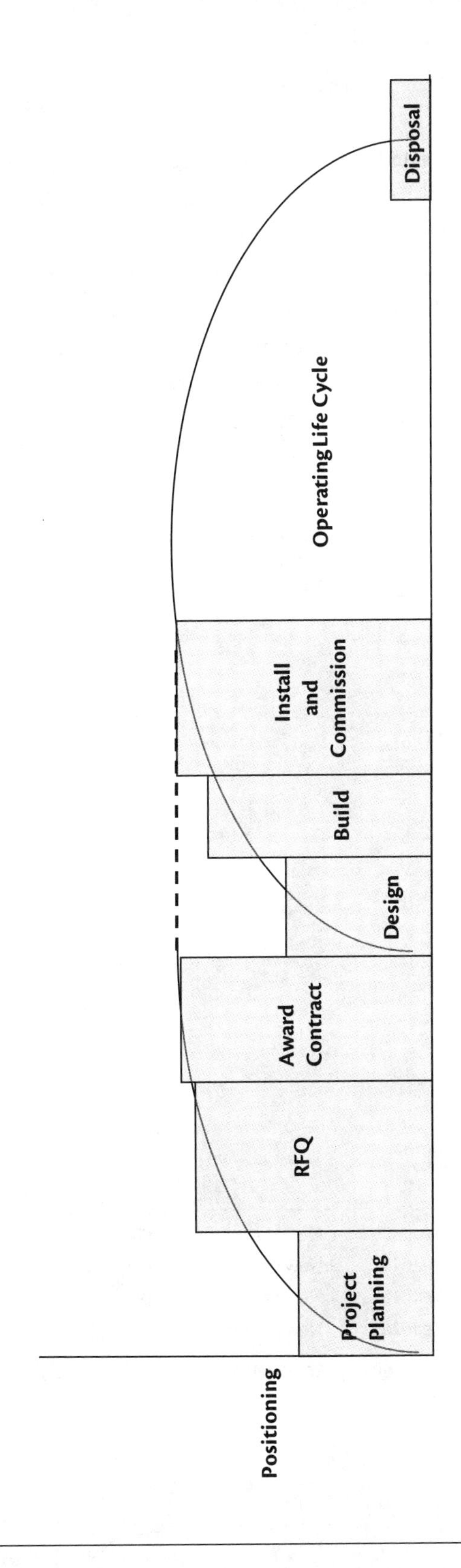

Figure 1-8 From project planning to asset operation to disposal (reprinted)

I want you to do something. Consider Figures 3-1 and 3-2 and give some thought to what information is being presented and what your response is to this question: "If this were your car, at what point would you consider selling it and buying a new car?" I just want you to consider each figure separately before moving to the next. Make a mark on the figure itself to indicate your decision point. It might be fun and revealing to have several people in your organization mark their 'decision points' to get a sense of the different interpretations of 'value' that exist among your associates. Start now.

Year of Ownership	**Purchase Price $30,000**	**Annual Payments**
1		$ 6,000
2		$ 6,000
3		$ 6,000
4		$ 6,000
5		$ 6,000
6		$ –
7		$ –
8		$ –
9		$ –
10		$ –
12		$ –
13		$ –
14		$ –
15		$ –
16		$ –
17		$ –
18		$ –
19		$ –
20		$ –

Year 5	New car price = $35,000
Year 10	New car price = $40,000
Year 15	New car price = $45,000
Year 20	New car price =$50,000

Figure 3-1 Mark what year you'd consider buying a new car

Year of Ownership	Purchase Price $30,000	Annual Payments	Fuel Mileage	Fuel Cost/Yr
1		$ 6,000	30 mpg	$ 720
2		$ 6,000	30 mpg	$ 720
3		$ 6,000	30 mpg	$ 720
4		$ 6,000	25 mpg	$ 800
5		$ 6,000	25 mpg	$ 800
6		$ –	25 mpg	$ 800
7		$ –	25 mpg	$ 800
8		$ –	22 mpg	$ 825
9		$ –	22 mpg	$ 825
10		$ –	22 mpg	$ 825
12		$ –	22 mpg	$ 825
13		$ –	22 mpg	$ 825
14		$ –	20 mpg	$ 850
15		$ –	20 mpg	$ 850
16		$ –	20 mpg	$ 850
17		$ –	20 mpg	$ 850
18		$ –	20 mpg	$ 850
19		$ –	18 mpg	$ 875
20		$ –	18 mpg	$ 875

Year 5	New car price = $35,000
Year 10	New car price = $40,000
Year 15	New car price = $45,000
Year 20	New car price =$50,000

Figure 3-2 Mark what year you'd consider buying a new car

What year did you consider buying a new car in Figure 3-1?

What year did you consider buying a new car in Figure 3-2?

If you chose different years, why? If you chose the same year, why?

Let's add more information in Figures 3-3 and 3-4.

Year of Ownership	Purchase Price $30,000	Annual Payments	Fuel Mileage	Fuel Cost/Yr	Mx Costs
1		$ 6,000	30 mpg	$ 720	$ 100
2		$ 6,000	30 mpg	$ 720	$ 100
3		$ 6,000	30 mpg	$ 720	$ 600
4		$ 6,000	25 mpg	$ 800	$ 150
5		$ 6,000	25 mpg	$ 800	$ 150
6		$ –	25 mpg	$ 800	$ 1,200
7		$ –	25 mpg	$ 800	$ 400
8		$ –	22 mpg	$ 825	$ 400
9		$ –	22 mpg	$ 825	$ 550
10		$ –	22 mpg	$ 825	$ 450
12		$ –	22 mpg	$ 825	$ 1,800
13		$ –	22 mpg	$ 825	$ 300
14		$ –	20 mpg	$ 850	$ 300
15		$ –	20 mpg	$ 850	$ 600
16		$ –	20 mpg	$ 850	$ 500
17		$ –	20 mpg	$ 850	$ 250
18		$ –	20 mpg	$ 850	$ 250
19		$ –	18 mpg	$ 875	$ 1,800
20		$ –	18 mpg	$ 875	$ 1,200

Year 5	New car price = $35,000
Year 10	New car price = $40,000
Year 15	New car price = $45,000
Year 20	New car price =$50,000

Figure 3-3 Mark what year you'd consider buying a new car

Did you choose a different year? Why or why not?

Year of Ownership	Purchase Price $30,000	Depreciation	Value of Car	Annual Payments	Fuel Milage	Fuel Cost/Yr	Mx Costs
1		20%	$ 24,000	$ 6,000	30 mpg	$ 720	$ 100
2		10%	$ 21,600	$ 6,000	30 mpg	$ 720	$ 100
3		10%	$ 19,440	$ 6,000	30 mpg	$ 720	$ 600
4		10%	$ 17,496	$ 6,000	25 mpg	$ 800	$ 150
5		10%	$ 15,747	$ 6,000	25 mpg	$ 800	$ 150
6		10%	$ 14,173	$ –	25 mpg	$ 800	$ 1,200
7		10%	$ 12,756	$ –	25 mpg	$ 800	$ 400
8		10%	$ 11,481	$ –	22 mpg	$ 825	$ 400
9		10%	$ 10,333	$ –	22 mpg	$ 825	$ 550
10		10%	$ 9,300	$ –	22 mpg	$ 825	$ 450
12		10%	$ 8,370	$ –	22 mpg	$ 825	$ 1,800
13		10%	$ 7,533	$ –	22 mpg	$ 825	$ 300
14		10%	$ 6,780	$ –	20 mpg	$ 850	$ 300
15		10%	$ 6,102	$ –	20 mpg	$ 850	$ 600
16		10%	$ 5,492	$ –	20 mpg	$ 850	$ 500
17		10%	$ 4,943	$ –	20 mpg	$ 850	$ 250
18		10%	$ 4,449	$ –	20 mpg	$ 850	$ 250
19		10%	$ 4,005	$ –	18 mpg	$ 875	$ 1,800
20		10%	$ 3,605	$ –	18 mpg	$ 875	$ 1,200

Year 5 New car price = $35,000
Year 10 New car price = $40,000
Year 15 New car price = $45,000
Year 20 New car price =$50,000

Figure 3-4 Mark what year you'd consider buying a new car

From Figure 3-1 to Figure 3-4, how did your answer change? Why did it change and what additional information did you have to make the determination?

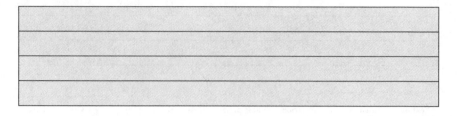

Do you recall the discussions in Chapter 2, specifically the phrases 'data-based' and 'decision-making criteria?' These are essential to making asset replacement or modification decisions. In the absence of data, and an unbiased and scientific way to makes asset decisions, we are just going to make a mark on a curve just like you did in the reprint of Figure 1-8 at the start of this chapter.

Let's look graphically at the data we saw previously. Figure 3-5 is the fuel mileage over a twenty-year projection that was shown in Figure 3-2. Note the trend line. What would be your acceptable risk level?

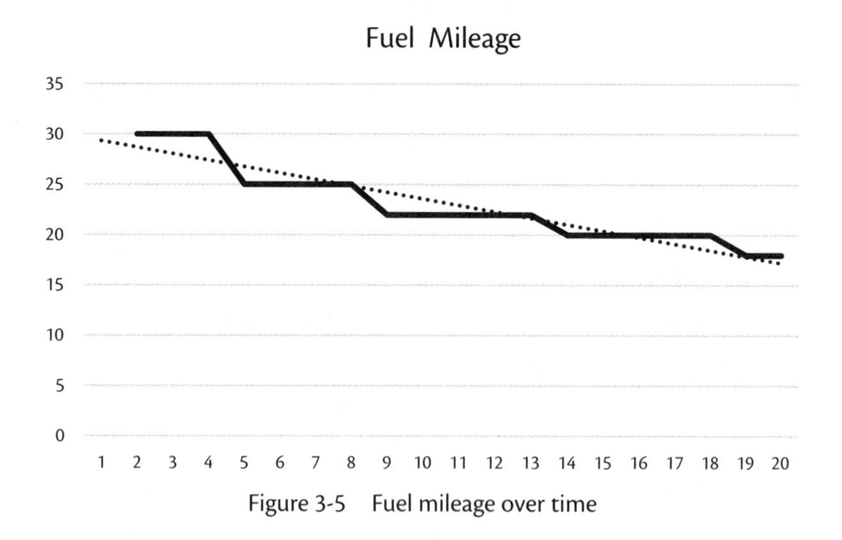

Figure 3-5 Fuel mileage over time

Refer back to Figure 3-2 and annotate here on Figure 3-5 where you had selected the year in which you would consider buying a new car. If you have had some other associates perform this exercise, mark their responses here on Figure 3-5 as well.

Did you notice that that your fuel performance was projected to drop 28% by year twenty? And that projected annual fuel costs were going to increase by 21%? And that the value of the car would drop 90% in twenty years?

Let's look at one more graph showing the car's depreciated value over time in Figure 3-6; again, note the trend line.

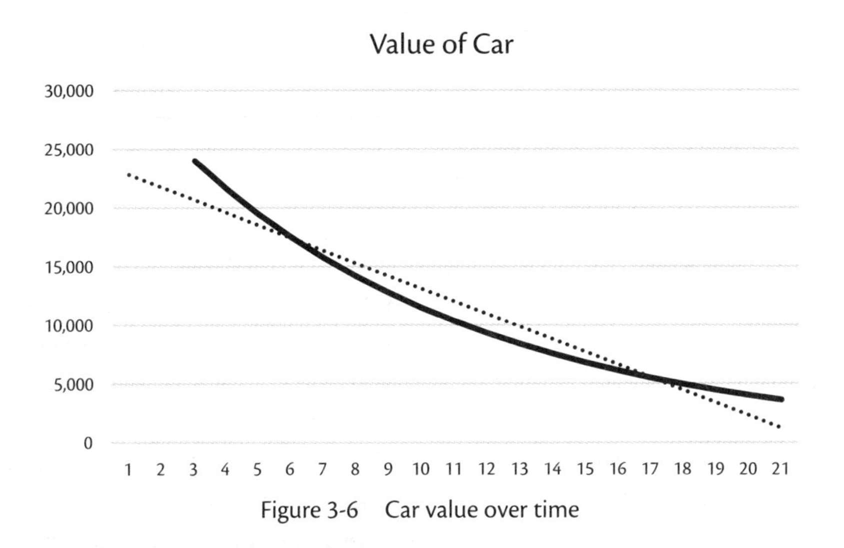

Figure 3-6 Car value over time

Indicate on Figure 3-6 the decision point you made on Figure 3-4. Have your associates record their marks as well.

Before we buy a "car," and I'm using air-quotes here, we should actually make the determination of when and how we are going to decide to buy a new car or modify a car for our upgraded purposes. For example, we might choose to hold on to an old car for the kids to drive, but we would very likely put some money into it to make sure it was safe and serviceable. This extra "kick" of money at the end would be reflective of the "disposal" cost shown in the reprint of Figure 1-8.

I used air-quotes earlier because I wanted you to mentally put 'asset' in that spot. What data do we collect and how are decisions made in your organization to replace or modify a machine? Shamefully and admittedly, here is the graph that I personally refer to when deciding to buy a new car or keep the old one. Figure 3-7 shows my personal shame.

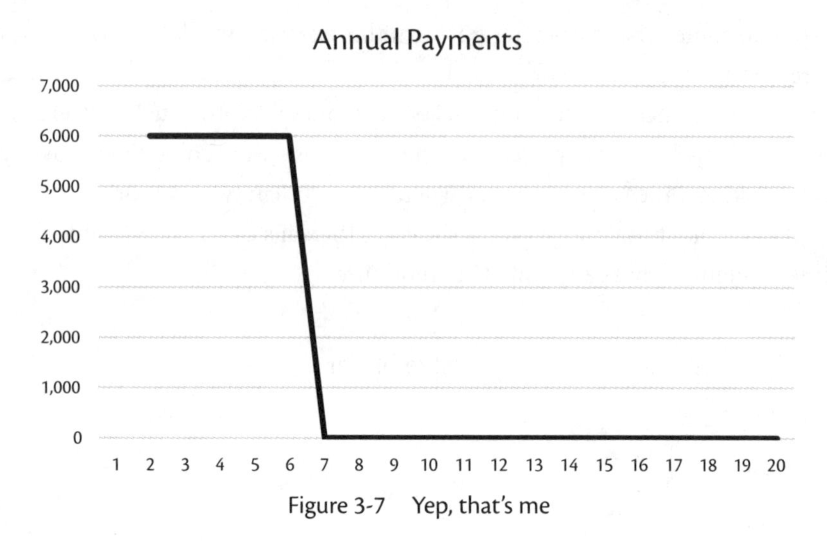

Figure 3-7 Yep, that's me

Be honest now. That's you too, isn't it? We often just hold on to an old piece of equipment in our facilities because it's already paid for and we think there are no other contributing factors adding to its diminishing returns for the organization. I'm willing to guess that you have an asset in your plant or facility that costs $250,000 brand new, and you've spent $2,500,000 maintaining it over the last fifteen years. When is it time to give up on the asset and buy a new one? The answer is in the data, and to curb our untethered biases, we have to pre-determine the value points where we make this decision. First, the required functions and needs of the asset must be agreed upon and well understood. As an example, you're likely to remove an asset that does not function as you need it to, or when the asset itself is no longer needed. How about removing it when we simply have paid all we intend to pay for the upkeep of the asset?

I keep coming back to the decision to sell a perfectly good Air Force staff car while it still had some value. Couple that thought with the fact that an asset's value does not necessarily end when we decide to get rid of it. Like a car, we want to maximize the trade-in value. Few can deny that a car has a better trade-in value when it is evident that the car was serviced regularly and correctly and well taken care of.

On the topic of modifying a machine, it is likely that most organizations make the rookie mistake of modifying machines to make them last 'just a little longer.' Or almost as bad, produce product faster. Modifying a machine for a different function is not in itself a bad move. Care should be given that the components that the asset is comprised of are not compromised on their performance curves with the new production venture. Almost all modifications for production improvement tend to reduce reliability and maintainability. This is a risky trade-off and runs counterculture to asset management.

I want to pull in a well-discussed point on equipment modification, specifically modifications for the purpose of getting more life out of the asset. This may be new to some of you, but see if this makes sense. Equipment arrives at our plant with an inherent level of reliability. You cannot maintain more reliability into a piece of equipment than it has inherently. To get greater reliability, you have to modify it for the purpose of increasing the reliability.

I used the phrase 'rookie mistake' to describe our equipment modification aims because it is a mistake I think we propagate all too often. We should modify equipment to last longer if it makes sense when comparing all the performance data (Figures 3-1 through 3-4). But the modification should be in this order: increase reliability, resulting in an increase in lifespan. It's likely, through our default 'production first' setting, that our order is: increase lifespan at the expense of reliability.

Maintainability and Reliability, or Having Your Cake and Eating it Too

I like cake, carrot cake to be precise. I really just wanted to say that. More to the point of this section, let's talk about how it is possible to build maintainability and reliability into an asset. By doing this, we are not only able to recognize more value from the asset (a tenet of ISO 55000), but have an asset that is more robust, easy to maintain, and will likely have low life cycle costs. Its like having our cake and eating it too.

First, a basis of understanding. For our purposes, maintainability is meant to be synonymous with Mean Time To Repair, or MTTR. This is shown graphically in Figure 3-8.

MTTR

| Response time | Troubleshooting time | Repair time | Startup Time |

Figure 3-8 Mean Time To Repair

Reliability is synonymous with Mean Time Between Failure, or MTBF, shown in Figure 3-9.

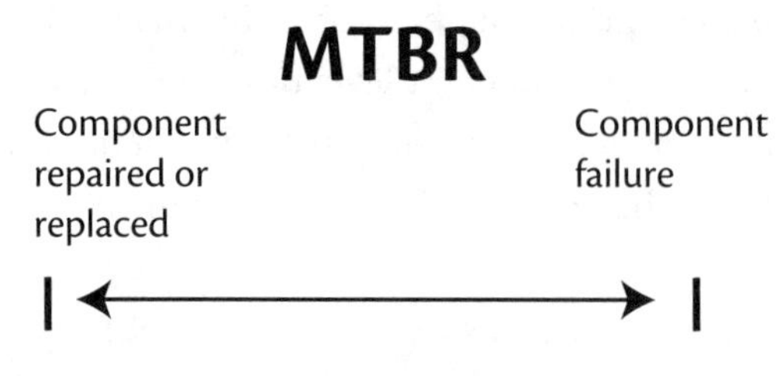

Figure 3-9 Mean Time Between Failure

As these are familiar terms to any reliability audience, there really is no need to go into any great detail on the calculations of these values. Instead, we will spend some time discovering ways to increase maintainability by *reducing* MTTR and to increase reliability by *increasing* MTBF.

ISO 55002 instructs a qualifying organization to have, as its plan, the rationale to perform the value preserving activities that it (the organization) is mandating. Furthermore, the plan is to include maintenance and operational necessities. Also, there must be measurements in place to determine if the plan is working

What is the rationale to decrease MTTR by increasing maintainability? The short answer is asset availability. If done right, this single element could also improve product output and overall productivity.

Reducing MTTR

Figure 3-8 is a rudimentary picture of what makes up the measure of MTTR. Let's break that down a bit further to determine *how* to improve maintainability:

- Response time
- Troubleshooting time
- Repair time
- Startup time

RESPONSE TIME

For the purpose of examining response time, as it relates to maintainability, take the angle of "how does an operator know there is a problem?" Is it possible for an issue to percolate under the surface and become a very large problem as it becomes more evident? Response time starts by first noticing that there is a problem. Can you list any design features that might be built into a new asset to bring issues of operational or performance concern to the attention of the operator?

Here are a few that I came up with:

- Match marking and other visual controls
- Witness marks (wear strips, cam followers, etc.)
- Alarms to indicate off-set point variables (temperatures, pressures, etc.)
- Sacrificial devices (shear pins, torque limiters, etc.)

Response time is further shortened by having straightforward communications between those in need (the operator) and those responding (maintenance). Think of first responders, like EMS and the fire department. I counsel organizations to consider forming a dedicated Do-It-Now Squad, or what I used to refer to as the 'Hit Crew.' A compelling argument could be made for this structure because ISO 55000 states that top management shall resource the organization with the necessary tools and people to carry out the asset management plans and activities.

The idea is to design an asset that has tattletale features to alarm the production crew of an issue. Clear and straightforward communication is then necessary to transmit the status of the situation, the scope and degree of this issue, and what aid is needed. With emerging technology, this is often an automatic task. Lastly, the responders need to respond quickly and be equipped with tools and skills. Or, as the ISO standards refer to it, with "awareness and competence."

Remember this critical detail for 'response time.' What features can be designed into the asset to shorten this time frame?

TROUBLESHOOTING TIME

There are several snowballing ideas to bring into this topic from the previous points on 'response time.' Namely, competence, awareness, and designing the asset for success. It would be reasonable to believe that a competent, well-resourced, dedicated team could have a dramatically positive affect on reducing troubleshooting time.

If you take *competent* to mean *skilled* and *knowledgeable*, you have the foundational idea behind such a response team. There has to be an element of 'design for maintainability' in an asset to reduce troubleshooting time even further.

What are some features you'd suggest building into an asset to make it easier to assess the situation and identify the offending component?

Here are a few that I came up with:

- Access doors
- Inspection windows
- Remote test ports
- Quick disconnects

- Positive locks and latches
- Match marking, and other visual controls

Remember this critical detail for 'troubleshooting time.' What features can be designed into the asset to shorten this time frame?

REPAIR TIME

Slightly relevant for 'response time,' more so for 'troubleshooting time,' but definitely a key attribute for 'repair time' is the idea of standardization of parts. This is most certainly a feature that should be designed into new assets and for all existing asset modifications. In fact, I feel so passionate about this practice that I would award it the highest level of mandate and say it *shall* be done (keeping with the theme of ISO 55001, of course).

Let's retrace a bit. Our measure is maintainability and our effort is to increase maintainability by decreasing Mean Time To Repair. So far, in our effort to decrease MTTR, we have:

- Made issues quicker to identify
- Improved direct requestor-to-responder communication
- Dedicated a competent and equipped response crew
- Designed into the asset the features to quickly assess the situation
- Ensured a high degree of standardization of parts

Coupled with some features identified in 'troubleshooting time,' (e.g., quick disconnects) how can the standardization of parts help with 'repair time'?

The next section will explore the idea of standardization of parts further; then Chapter 5 is a protracted discussion of the Bill of Materials, which will again touch on this subject.

For now, remember this critical detail for 'repair time.' What features can be designed into the asset to shorten this time frame?

START-UP TIME

If you have attended one of my classes, or I've ever consulted in your facility, you would have heard me describe the perfect maintenance job. Let me describe a planned maintenance job in seventeen steps:

1. Operator is aware of scheduled maintenance; machine is still running.
2. Maintenance technician arrives on time, with materials and equipment.
3. Maintenance technician and operator discuss other issues that might be present.
4. If there are other issues, the maintenance technician will call his or her supervisor to the machine.
5. The maintenance technician asks the operator to keep the equipment running while the maintenance technician walks around the asset performing a look, listen, and feel inspection.
6. The maintenance technician will alert the operator and the maintenance technician's supervisor if other concerns are detected.
7. The maintenance technician instructs the operator to clear out the machine, and shut it down in a controlled manner.
8. The maintenance technician locks out the machine.
9. The maintenance work that was scheduled is executed.
10. The maintenance technician removes the lock out.
11. The operator restarts the machine while the maintenance technician is still present.
12. The operator runs five 'good' parts (or the equivalent).
13. The operator confirms proper operation.
14. The operator signs a survey on the back of the work order indicating his or her level of satisfaction with the maintenance that was accomplished.
15. The maintenance technician signs a survey on the back of the work order indicating his or her level of satisfaction on how ready the asset and the operator were for the scheduled work.

16. The maintenance technician completes the work order documentation.
17. The job is done.

Steps 10-13 are the 'start-up time' steps. Regardless of the nature of the repair (planned, unplanned) the start-up steps are the same.

We've had a common theme running through the first three MTTR elements: "What features can be designed into the asset to shorten this time frame?" That question is a little more puzzling with start-up time.

Give some consideration to what you might suggest if you were on an asset design team, or an asset modification design team. Further, recall that ISO 55000 sets an expectation that organizations are going to work to identify and correct issues and to capitalize on opportunities. How might you suggest start-up time could be reduced? You may have to call in some distant learning on Failure Modes and Effects Analysis (FMEA) or Reliability Centered Maintenance (RCM).

I have a few thoughts on this matter. It might start by ensuring the operator is present to start the machine and that there is no other reason to 'not start the machine.' That might facilitate the design and practical feature that the asset is indeed ready to be started. This sounds ridiculous on its face, but hear me out. Imagine a pump that has to be primed, or a diesel motor that won't start because it's too cold, or power to the big motor is restricted due to some electricity curtailment. These can lengthen (the opposite direction we want to go) the start-up time.

We have discussed in some detail how to decrease the Mean Time To Repair and thus improve maintainability, but how about reliability? How can we increase reliability?

Increasing MTBF

Take a quick look back at Figure 3-9. It's ok, I'll wait. There really isn't much to this image. At the most basic level we are measuring how long an asset runs between failures. In an even more drilled-down view we want to know how long each component will last between failures.

MTBF itself is not a complicated measurement, but what muddies the water is defining exactly what a failure is. One of the companies I work with tracks Mean Time Between Breakdowns; I believe this to be a similar measure. Seriously though, what is a failure or a breakdown?

If you really thought hard about it, and brought in the generally well accepted study of something as vast as Overall Equipment Effectiveness (OEE), you might think you were doing some good. But honestly, in OEE, what is the dividing line between the Availability sub-element of breakdowns, and the Performance sub-element of minor stoppages? I'm not sure of the answer, but I do know in all instances of these two occurrences, no product is being made, and thus there is no value to the organization.

Here's a scenario for us to consider to make this asset-to-component analogy. I go back many times to drawing a comparison with vehicle maintenance because I think everyone, regardless of their station in the organization, can relate to this type of storytelling.

At the time I am penning this book, I own a 2011 Ford Escape. Before two-thirds of you roll your eyes, understand that my car is required to get me to the airport and back. I'm not a gearhead, I just need a car for transportation and I happen to like Fords.

My car, as an asset, has a Mean Time Between Failure of 400 miles. The limiting component is the gas tank. My car gets between 26–27 mpg and has a 15-gallon tank. I never run my car to failure for gas. I usually refuel when the idiot light comes on showing that it's time to get some gas. This light indicates that I have about 50 miles of typical driving left before a 'failure' or the failure mode of: engine stalls due to lack of fuel.

Let's build this thought:

- Asset (car) MTBF is 400 miles
- Limiting component (sub-asset) is the fuel tank, which holds 15 gallons
- Asset (car) performance is 26–27 mpg
- Asset management plan is to refuel every 350 miles
- SAMP elements are to always operate the car within range of a gas station based on how much gas I have in the tank. Also, keep engine RPMs low (that burns the gas)

I have stressed the point that I don't run out of gas, and explained my technique of refueling when the light comes on. Since my car does not experience this failure mode, the next limiting component on my vehicle is the oil. I did not specify the oil filter because I usually change this at the same time as I change my oil. I use five quarts of synthetic oil, brand and viscosity as recommended by the Ford Motor Company. I change my oil every 5,000 miles. It is important to note that synthetic oil does not 'fail' at 5,000 miles. In fact, most synthetic oil can be used up to 10,000 miles and some even beyond. This is a driver/owner preference. I reduce my risk by controlling the probability that I'll have a problem. Remember: Risk = Probability × Consequence. The consequence would be that I am stuck on the road with a seized engine.

- Asset (car) MTBF is 5,000 miles
- Limiting component (sub-asset) is the oil
- Asset (car) performance mandate is no oil leaks, no burning of oil, no filtration issues
- Asset management plan is to change the oil and filter every 5,000 miles at a Ford dealership
- SAMP element is to use manufacturer recommended oil and filter, do not exceed 5,000 miles + 10%

My car never fails because of an oil issue, or the failure mode of: engine seizes due to lack of proper lubrication.

The next limiting component on my car is the tires. People think I made this number up, but, at the writing of this book, I have 160,800 miles on my car. I got 90,000 miles on the stock tires, and I'm on my second set and look to get the same performance. I rotate my tires every 5,000 miles, and rotate and balance every 10,000 miles. I do not service (inflate) my tires outside of this schedule. I would like to say it's because I'm too busy, but it's really because I'm too lazy. I have a full-size spare tire.

- Asset (car) MTBF is 90,000 miles
- Limiting component (sub-asset) is the tires
- Asset (car) performance mandate is no issues or uneven wear on any tire

- Asset management plan is to rotate tires every 5,000 miles and rotate and balance every 10,000 miles, inflate to proper pressure during each service
- SAMP element is to use manufacturer recommended tires, operate vehicle properly, avoid potholes and road debris

I have run over a few nails in my day, but I've avoided all the debris I could see.

This analogy could go on for many more components. I usually sketch out a few for my classes: serpentine belt, spark plugs, etc.

As an example for your location, list that asset you initially recorded in Table 1-1:

This asset has an MTBF. It is made up of several components, five of which are:

1.
2.
3.
4.
5.

The important takeaway from this discussion on increasing MTBF is to identify the MTBF of the asset, and then break that down and see what the MTBF limiting components are. I often tell clients that one of the primary roles of a Reliability Engineer is to know these components and know which components are responsible for limiting the MTBF. From there, the Reliability Engineer should be tasked to form a small team, study the data, and devise a plan for improving the MTBF. This is a very simple and continuous process.

When we discussed improving maintainability, we touched on the importance of standardization of parts. For MTBF, what would be some benefits you could suggest for adopting a process of parts standardization?

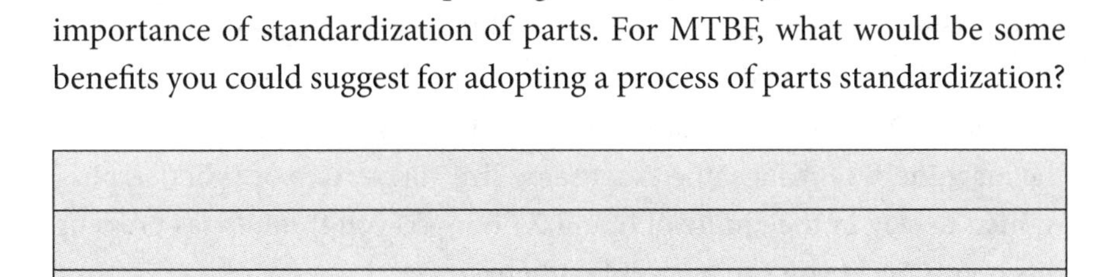

We are going to continue that thought right after this short discussion on why any of this matters.

Why Do MTBF and MTTR Matter?

I promised a short discussion on why Mean Time Between Failure and Mean Time To Repair matter. Well, here it is. In Figure 3-10, Inherent Availability is a product of only MTBF and MTTR.

$$A_i = MTBF/(MTBF + MTTR)$$

Figure 3-10 Inherent availability

The further we drive MTTR to zero, the closer we get to Ai = MTBF. That would be an ideal state.

Standardization of Parts

I'm going back, briefly, to a concept introduced in Chapter 2: the Strategic Asset Management Plan (SAMP). I shared some thoughts and definitions on what the SAMP was, and I gave an example of a game of chess as a way to think strategically. Always stay several steps in front of your opponent. This is a good analogy.

Let me share this story. Years ago, I was a central player on a design team for the construction of a new manufacturing plant and for all the equipment to be built for that facility. In an engineering meeting with the general contractor discussing equipment components, I recommended a brand name for hydraulic cylinders. I told my fellow engineer (the contractor) that I wanted all the equipment that needed hydraulic cylinders to have brand X cylinders. I wanted parts standardization. I realized that all the machinery can't have the exact same size and service of cylinder, but I wanted to stay in the family of brand X. Why do you think I was pressing this point? See if you can answer that here:

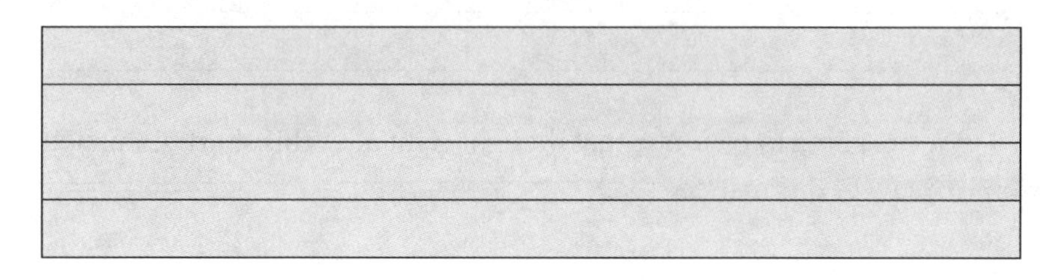

My contractor asked me why I would want to do that. I didn't initially have an answer. I was so dumbstruck that a fellow engineer would not know the value of such a request and instantly know that I was going for standardization of parts. More specifically, he should have known that I was recommending the most truly vetted and robust parts. The hydraulic cylinders I was recommending were the ones that my company already used in other similar manufacturing facilities and proved to be the most robust for the crazy environment we were running in.

I submit to you that if we don't center on a list of parts that we are going to standardize around, then we are going to end up with three different types of PLCs in our plant, and seven flavors of variable frequency drives. Standardization issues always seem to be the greatest around electronic (4-20 mA) components.

The failure pattern for electronic parts is the traditional early-stage failure or what we used to call 'infant mortality.' You can't see this type of failure coming; it always results in a breakdown. The only strategy we have to work with is to have an exact spare replacement. Why? Because you never know when an electronic component is going to fail. It fails randomly (failure pattern) and suddenly (deterioration curve).

Standardized parts are known to our technicians and to our operators. Our company can share ideas, and standardized spare parts across the country or globe. Certainly in the twenty-first century we should be able to tell an OEM (Original Equipment Manufacturer) what components we are going to accept on our assets.

We can then work to develop the best way to perform preventive maintenance on brand X cylinder known to mankind!

I've danced around this for a while, but I want you to list some benefits of having a standardized list of parts for your facilities:

Now, let's increase the level of difficulty with this next question. Thinking strategically, how would the idea of standardized parts help your organization? Again, remember the SAMP. Tie spare parts standardization to the availability of the asset to generate value for the company five, ten, or twenty years into the future. I'm going to give you some space to make a compelling argument:

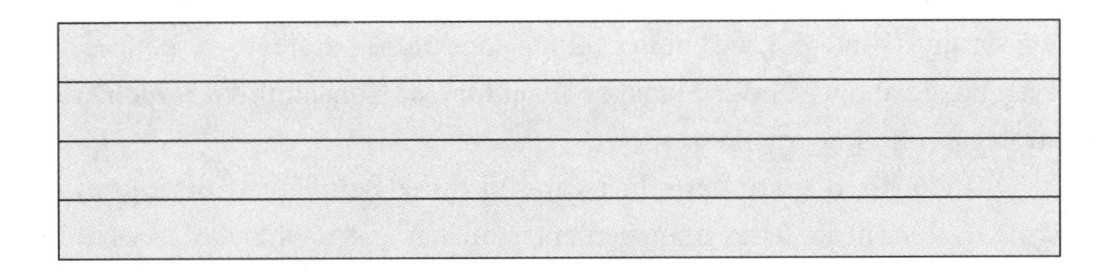

It is through the standardization of parts that a company can actually start to flex some muscle and demand better service and pricing. Consider why third-party storeroom contractors and major parts suppliers get such a better deal on parts. It is because they often deal directly with the manufacturer of those parts. You might buy two brand X cylinders every six months, but your favorite parts supplier buys 1,000 every month. Yet, you might have five different brands of hydraulic cylinder at your site and are unable to gain any leverage on buying.

It's a little known fact, but Deming actually advocated putting all your eggs in one basket. Most purchasing department managers are insistent that we have more than one vendor for each category of parts. In fact, it might be a common belief that to do otherwise could invite a supplier to take advantage of your company.

Deming pointed out that you should limit your suppliers. In that way, you can leverage a better price and better technical service.

Here is an example that I like to share:

- Billy's motor supply company has 35% of my business
- Sally's motor supply has 45% of my business
- Jim's motor supply has 20% of my business

If I ask Billy for a price break on motors, he is likely going to say there's not much he can do without getting more of my motor business. Jim would certainly say the same. Maybe even Sally. But, if I gave any of them 100% of my motor business predicated on the delivery of a much better price and free engineering application advice, then I might just get it. I'd always throw into the deal an insistence on having 100% return to supplier ability at $0 cost.

With standardized parts, I can think strategically. I just mentioned how we could get a better price on future purchases (strategic), better engineering service (strategic), and better return conditions (strategic). We haven't even talked about Vendor Managed Inventory, or Consignment, which are strategic sourcing practices.

Standardized spare parts helps me to think big picture, on a grand scale. My strategic asset management plan has got to include access to spare parts in the future.

Consider this analogy to having the 'stuff' you need to do your job. Think strategically.

A company that sells lumber needs lumber. That company is likely to own the lumberyard. In addition, they most likely own or have controlling interest in the saw mill. They would be foolish to be at the mercy of someone else's schedule for lumber production. That same company is likely to own the forest. The company might also own the lumberjack services for harvesting the trees. This is integrated supply. This is strategic.

I imagine that if you're in the corn products business and your company doesn't own the corn fields, then your company has corn futures locked up for ninety-nine years. That's strategic.

Sometimes it's helpful to drive a point home by building a strawman argument against the proposed position. Here's such an argument. It will sound ridiculous, but I believe it has some real merit.

Let's say that your company has decided *against* compiling a comprehensive list of spare parts, makes, models, brands, etc. Maybe they haven't decided against this effort, but rather they haven't gotten around to it yet. As a result, your company does not have an insistence on what parts are acceptable to be built into every new asset, or every asset to be modified. If, on the other hand, you do have such a list of standardized parts, then current suppliers can be notified that these are the components you are going to stock and order. OEMs can be put on notice that, "if you want to sell a machine to my company, here are my rules."

Without this purposefully created list of acceptable parts, your company, by omission, has abdicated what parts go on to future equipment to the OEM, so the OEM can use whatever parts they get the best deal on. In fact, your company has, as a business decision, decided to roll the dice and let fate lead them where it may. The OEM is thinking strategically, why aren't *you*?

Does any of that sound like a process or policy that you would actually suggest be adopted? In a very large sense, isn't this where we actually are?

Standardization of parts is actually an intermediate parts process incorporated to make a storeroom more efficient. Examine Figure 3-11 to see exactly where parts standardization fits in the storeroom improvement process. Figure 3-11 is a roadmap a colleague of mine and I devised many years ago. This roadmap is the actual tool I used to illustrate where an organization's storeroom processes are presently, and to help them develop a strategic and tactical route to world class.

As a strategic position, parts standardization is one of the most significant steps we (as leaders) can take to make assets manageable. If you will agree that all our assets are made up of components, or parts, then that is the first premise of this belief. Agree with me then, that all our assets are made of up the *same* types of components. Every machine in our plants or facilities is made up of these 'kinds' of components:

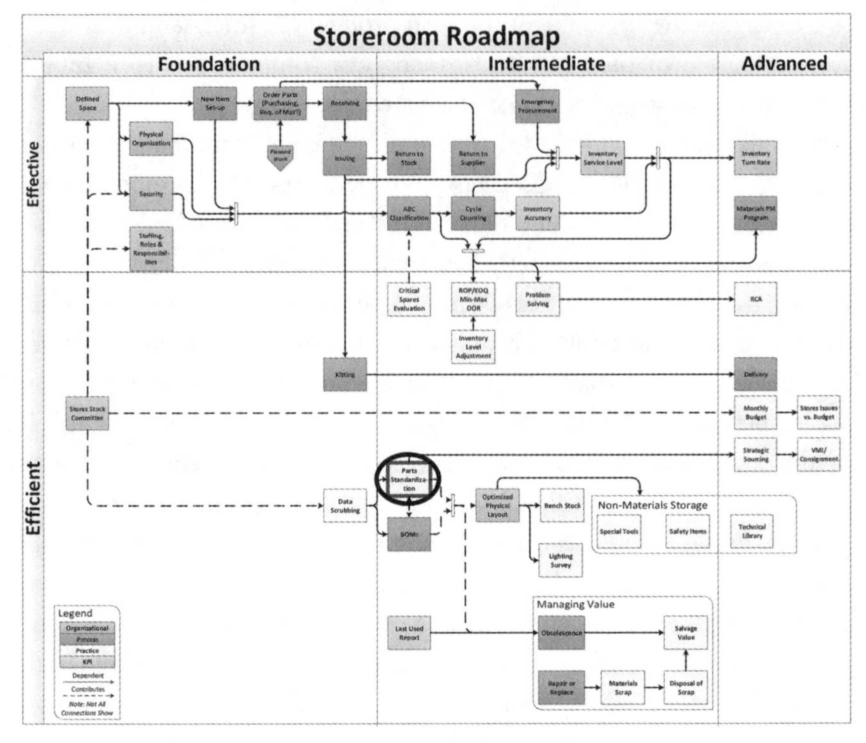

Figure 3-11 Parts standardization on the storeroom roadmap

Adapted with permission by Marshall Institute, Inc.

- PLC
- Valve
- Switch
- Sprocket, chain, sheave, belt
- Wiring harness
- Motor
- Coupling
- Shaft
- Gearbox
- Cylinder

If we can manage (read that as 'control') what brands and makes of components are allowed on our equipment, then we can agree on the best and most robust parts to use in our operating environment. Or, to use an ISO 55000 term, *operating context*. Further, and as demonstrated, we can get better pricing and aftermarket service. We also have a much better opportunity to leverage opportunities gained by the analysis of our Reliability Engineers when they are examining limitations to component MTBF performance.

I'm going to leave you with a final thought on parts standardization. Most reading this book will have consumed a carbonated soda at some time in their lives. With the health trends as they are in the 21st century, I would understand if many of you may not drink soda on a regular basis, but at least you may have had a favorite at one time.

Here's the last thought. Have you ever gone to a restaurant and asked for Coke and the server asked if Pepsi was ok? If you are a Coke drinker, the choice of Pepsi was not ok.

If something as insignificant as the *wrong* brand of soft drink can bring a response of dislike to our minds, why are we more forgiving of an 'off spec' spare part that literally goes on the asset that creates our paycheck? We can't control what beverages the restaurant sells, we are at their mercy (by the way, they serve what they get the best deal on, like OEMs). But we certainly can, and must, control spare parts through standardization.

After we're done installing or modifying the assets, we're all done. Right? Wrong.

Don't Quit in the Middle

Oh, we are not done after a new asset is landed, commissioned and running, nor when a machine is fully modified and back in service. No, no, no. We're just in the middle.

We talked at great length in Chapter 2 about the SAMP. With a new asset, or a modified asset, we are just at the front end of creating or changing the plan. The asset management plan includes the 'stuff' we do to maintain the asset to ensure it provides value for the organization. This value is more than likely a projected value. After all, new assets are just at the beginning of their operational phase, and modified machines are just ramping back up in hopes of producing at a greater volume or speed.

I can suspend with the pretense here and concede the point that no companies out there actually have a full complement of spare parts in the inventory before they put a new or modified asset into full production. Nor is it likely that any companies have the newly minted PM protocol in the CMMS prior to firing up the new asset. With this most likely the case, the work on PMs and spare parts has yet to be done.

Let's address the spare parts needs first. A new asset that is being installed by a contractor for the OEM, or by the OEM, will arrive with some commissioning spares. These are some well-vetted items that the OEM has experienced issues with during previous start-ups. In fact, they are very likely to be called 'commissioning spares', and some of them will be used to get the machine up and running, tested, and verified. These are not operational spares.

The difference between commissioning spares and operational spares is simply when they are used. Commissioning spares are used in the commission of an asset, and the operational spares are the ones that will be used throughout the asset's operation at your location unless otherwise superseded.

There is no value to an OEM when they ship commissioning spares back to their shop. As such, they are almost always willing to sell them to you for pennies on the dollar. And you will take them. After all, you need spare parts, right? You may find that you will hold on to these commissioning spares forever and they will soon become obsolete parts in your storeroom taking up valuable space, time, and oxygen.

The operational spare parts are the actual parts you need to keep the equipment serviced and functioning for years to come. The OEM reluctantly gave you a maintenance manual and a poorly detailed call out on the Bill of Materials. Along with that offering, the OEM issued you a list of recommended spare parts, with their (the OEM's) part number. They do all this in hopes that you will buy the spare parts from them.

Everyone reading this book will contact their favorite parts supplier, name here,

and ask them to give you the details on the parts and provide a cost proposal for them. I know you do it. I did it myself for decades.

The fact is, we need this list of spare parts to stock the storeroom. We aren't going to stock everything on the Bill of Materials (BOM), but we are only going to stock items that appear on an asset's BOM. This is part of our asset management plan.

Preventive Maintenance is another aspect of operating a new asset that is every bit as critical as the spare parts. Although each asset in our plants is generally made up of the same components, each of those components is running in a different environmental and operational context.

For example, you might have a simple 5-hp motor running a pump in a sump area in the basement. The conditions are cool, damp, and musty and the assembly has a limited run cycle, only running when necessary. That same motor might be driving an exhaust fan in the attic region of a silo and it never stops. The conditions are hot, dry, and dusty. The PMs should be different for each context of operation and environment.

There are three reasons we can't seem to get our initial PMs correct for new equipment or modified equipment:

1. We don't know better.
2. We don't know how.
3. We're lazy.

I sure hope it's the second excuse.

Do not copy and paste PMs from one asset to another. The only time this is permissible is if the two assets are exactly the same, performing the same service and function, for the same duty time, and in the same context. That is a lot of qualifiers.

The spare part strategy and the preventive maintenance strategy for each asset in our facility is part of the inherent makeup of the asset management plan. These are integral elements of the stuff that we actually do. In fact, preserving the inherent reliability of an asset is contingent upon the practice of like-for-like parts replacement and executing preventive maintenance steps as prescribed.

Thinking strategically, we have to adopt the mandated requirement that PMs must be drafted prior to starting the machines in our production effort, and that we have on hand the necessary operational spares and a strategy to keep both current and relevant.

Take the opposite approach and see if it makes sense. We plan not to have a preventive maintenance strategy on the new asset until we are about six months into production. Even then, we will just copy and paste it from a machine that is similar. On the subject of spare parts we have a 'wait and see what breaks' philosophy to determine what parts we actually need to stock.

Everything mentioned in this section applies not only to new assets, but rebuilt and modified equipment as well, so don't forget those assets in our facilities.

Chapter Summary

Chapter 3 focused on capital assets, which is the focus of this workbook regarding the application of asset management. Our journey began by determining whether a new or modified asset was best for our needs. In either case, new or modified, we should design for maintainability and for reliability. Key to our success with the strategic asset management plan was the idea of parts standardization. We further recognized that when the new machine was up and running or an asset was modified and back in service, our work was not done.

Adding to Our Business Case

We need data to make asset decisions. These decisions center on whether it is in the best interest and in alignment with organizational objectives to purchase a new asset or modify an existing asset. This is in concert with the ISO 55000 requirement to be data-based, and have decision-making criteria.

An analogy to this thought is one where consideration is given to purchasing a new car. An example of the data needed to make a new car decision might include:

When purchasing a new asset or modifying an existing machine, maintainability and reliability should be initial elements for consideration and overall requirements.

ISO 55002 instructs a qualifying organization to have, as its plan, the rationale to perform the value-preserving activities that it (the organization) is mandating. Furthermore, the plan is to include maintenance and operational necessities. There must also be measurements in place to determine if the plan is working. The activities involved in improving maintainability and reliability are the 'preserving activities' referenced.

The rationale to decrease MTTR by increasing maintainability is an increase to asset availability. If done right, this single element could also improve product output and overall productivity.

Maintainability is synonymous with MTTR and is generally measured in four distinct elements:

1. Response time, meaning:

2. Troubleshooting time, meaning:

3. Repair time, meaning:

4. Startup time, meaning:

To decrease an asset's MTTR, we need to shorten one or several of these four elements. As an example, we can do this by:

The rationale to improve reliability is also an increase to asset availability. Reliability is synonymous with MTBF and is commonly considered the time between an asset failing, or breaking down, and the asset being brought back up into production. Not only does the asset have a MTBF value, but the components on the assets have their individual MTBF values as well.

As an example, our current asset:

has an MTBF. That asset is made up of several components, five of which are:

1.
2.
3.
4.
5.

These five individual components each have their own MTBF. To increase an asset's MTBF we need to understand what components are contributing to the asset's reduced MTBF.

One of the chief ways to understand each component is to adopt a philosophy of parts standardization. Parts standardization contributes to the strategic asset management plan by:

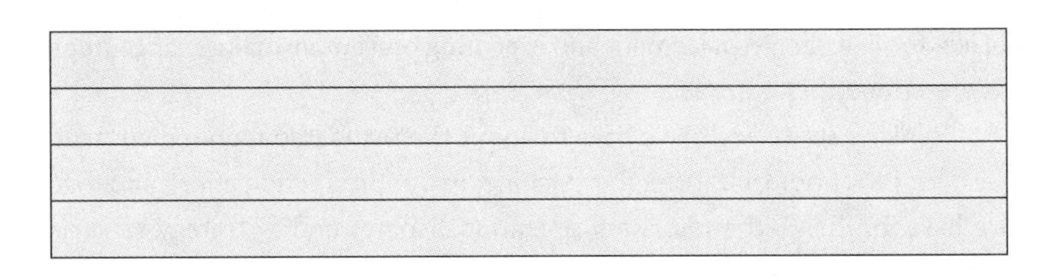

As mentioned, an important necessity to improving MTBF is to identify the MTBF of the asset, and then break that down and see what the MTBF limiting components are. Our job is to know these components and know which components are responsible for limiting the MTBF. From there, small teams will study the data and devise plans for improving the Mean Time Between Failure. This is a very simple and continuous process.

MTBF and MTTR matter to our business because these two reliability metrics make up the Inherent Availability measure as shown in the following formula:

$$A_i = MTBF/(MTBF + MTTR)$$

The further we drive MTTR to zero, the closer we get to Ai = MTBF. That would be an ideal state.

After assets are purchased and installed or modified the work is not done. If not completed prior to the asset startup, PMs need to be modified or created, and a comprehensive listing of spare parts needing to be stocked will require the storeroom's engagement. Spare parts and the preventive maintenance protocol on assets square with the mandates of the asset management plan and the SAMP. These actions are the 'activities' referred to in the ISO standards. They are literally the 'stuff we do.'

Preventive Maintenance is another aspect of operating a new asset that is every bit as critical as the spare parts. Although each asset in our plant is generally made up of the same components, each of those components is running in a different environmental and operational context.

The spare parts strategy and the preventive maintenance strategy for each asset in our facility is part of the inherent makeup of the asset management plan. These are integral elements of the activities that we actually do. In fact, preserving the inherent reliability of an asset is contingent upon the practice of like-for-like parts replacement and executing preventive maintenance steps as prescribed.

Thinking strategically, we have to adopt the mandated requirement that we draft PMs prior to starting the machines in our production effort, and that we have, on hand, the necessary operational spares and a strategy to keep both current and relevant.

Turn to the last section of the book and record this same information in the comprehensive strategy section.

FOUR

BOMs—Parts of the Whole

"One of the more important and complex facilities in any asset maintenance operation is an adequately stocked and controlled spares and materials store. The level of complexity can vary from the difficulties experienced in catering for breakdowns to the easier situations being brought about as predictive and preventive maintenance regimes are introduced."

—Dr. Alan Wilson, *Asset Maintenance Management* (p. 672)

This chapter is intended to cover more than just the Bill of Materials (BOMs). It is my hope to conduct a deep dive into spare parts and the storeroom itself. No discussion on spare parts or the storeroom can be complete without an understanding of the BOMs. In fact, I'd go as far as to say that no discussion on the storeroom or spare parts can even *begin* without an agreement on BOMs. This chapter will provide a glimpse into this most critical element of asset management by providing a definition for BOMs, a clear link to who is responsible for providing the BOMs, how to use BOMs to determine what parts to stock, and different stocking methods for spares; the chapter will end with a thorough discussion of the storeroom's role in asset management.

The ISO 55000 series of documents is riddled with pronouncements that the organization must have systems and processes in place to measure, evaluate, and essentially deal with risks of all kinds. Who can deny that the all-consuming, time-intensive process of locating and procuring the necessary spare parts is putting all of industry at risk?

One necessary element, buried in ISO 55002 to be specific, is a section on management of change, with a small note that the organization should address supply chain constraints as a means to avoid an increased leveraged risk by this particular choke point (that being the lack of access to spare parts). This is rationale enough to address, at the highest level, what most of us have known for a very long time. Namely, that spare parts or the dearth of them, are our Achilles heel!

What's troubling, especially on this subject, is the absolute avoidance of the truth that our industry of reliability and maintenance has allowed to occur. We have allowed for an asset's BOM and the spare parts required to keep an asset operational to become two separate discussions. What this might look like in real life is having spare parts in stock in the storeroom that don't actually belong to any asset in the plant. Every organization has them. In fact, there is a world-class level for obsolete parts; there should be <5% (by value) of obsolete spare parts in our inventory. That means one thing if your inventory value is $400,000 and quite another if it is $3,000,000.

So what, right?

So What's the Big Deal?

It's safe to ask what the big deal is about having parts in the storeroom that don't go to anything. In fact, this surplus inventory gives our maintenance crews a fighting chance when they don't have the right part, but they can make *something* work. I think I saw a MacGyver episode one time where Angus (that is actually his first name) was attempting to bring a production plant back on line with only a paper clip and a chewing gum wrapper. Well, since it was a maintenance job, there must have been a roll of duct tape used as well.

Why don't we identify the spare parts that don't go to anything and get rid of them? Instead we can begin to stock parts that are actually associated with our current equipment.

Obsolete spare parts give us a false sense of security, allowing us to think that we have plenty of stock to provide for any job. However, there is a carrying cost component to running an effective and efficient storeroom, and that cost can exceed 42% in rare cases.

What is more common is to have a carrying cost (the cost to stock something) that hovers around 25–33%. Just to put that in a reference that everyone can understand, if you have a storeroom with a carrying cost of 25%, every four years that the item sits on the shelf you have just *purchased* it again (.25 × 4 = 1.00). For a carrying cost of 33%, it drops down to three years. That is real money!

What values are used to calculate this carrying costs? Following is a list of variables that I traditionally use to arrive at a *real value* that a client can use to calculate their carry costs:

- Square footage of the storeroom property
- Cost of a square foot of warehouse rental space in the local community
- Wattage of lamps in the storeroom
- Costs of a kWh of electricity for the client
- Inventory value
- Current interest rates on a savings account
- Storeroom labor costs (grossed up for benefits)
- Insurance premium on any stock
- Annual tax on any stock

Figure 4-1 is a sample of the results of a storeroom savings calculation. About four years ago, I invented a Potential Storeroom Savings Calculator. I use this tool to articulate potential savings that could be reached if a storeroom operation decided to really embrace world-class principles. My intention, and the use of this tool, has been to aid clients by providing the savings potential data and the carrying costs so they could make a compelling case for change to senior management.

	Current	Ideal
Inventory Value	$ 3,575,000	$ 1,000,000
Carrying Cost	$ 738,080	$ 206,456
Total (realized)	$ 4,313,080	$ 1,206,456
Potential Savings		$ 3,106,624
Carrying Cost (as a percentage)		20.65%

Figure 4-1 Potential savings calculation results

In Figure 4-1, the client had a current inventory value of $3,575,000 and a carrying cost of 20.65% (calculated by the tool). Their carrying cost on a value just north of $3,500,000 is $738,080. That literally means that this plant spends almost $740K each year to keep a stock of $3,575,000 annually! This is added to the actual inventory for a total 'realized cost' of $4,313,080.

This particular plant has an Estimated Replacement Value of $100,000,000. The calculator tool arrived at an *ideal stock value* of $1,000,000. At 20.65% carrying cost, the carrying cost would be lowered to just over $206,000 if the inventory value was lowered to the ideal sum. The net is a potential for savings of over $3,000,000. Remember from the previous point, at 20% carrying costs, this company literally 'buys' their inventory again every five years!

Give Figure 4-1 another look and consider this startling fact. A world-class storeroom might have up to 5% of the value of its inventory as 'obsolete.' In Figure 4-1, under the *ideal* inventory value, they could still expect to have up to $50,000 in obsolete items in the storeroom at any given time. This particular client could be sitting on $178,750 of obsolete inventory (at 5%) with their current inventory value, and paying annually for the privilege!

What I didn't share with you before, but I will now, is that a typical storeroom might have 20–30% of its inventory as obsolete. Again, those are items that don't go to anything in the plant.

If you were reaching for your calculator, let me help you out. My client had the potential to have an obsolete inventory value of $1,072,500 and pay a carrying cost on top of that.

So, what's the big deal? The big deal is that our maintenance and storeroom personnel dig through obsolete parts to find the right part. It's like a treasure hunt. Imagine a maintenance technician on the midnight shift putting a valve with a brass seal on the hydrochloric acid line versus the valve with the ceramic seal. The result of that misstep might be the Channel 9 News helicopter flying over your plant. There is a real risk to stocking items that don't go to any asset.

A part is considered obsolete for an asset if it does not belong on that asset. A part is considered obsolete in our plant if it doesn't go to anything in the plant and is not on the BOM for any asset.

In the Scheme of Things

In the last chapter, I shared a road map that was created years ago. I still use it to establish where an organization's storeroom is operationally, in the scheme of things. It is of strategic importance to clearly establish where the storeroom is compared to where they should be, performance and capabilities-wise. As a colleague once told me, "Even if you know where you're going, you're still lost if you don't know where you are." Knowing the starting point, or "where you are," is fundamentally important in determining what route to take on a map. Figure 4-2 is the storeroom roadmap indicating where BOM is addressed.

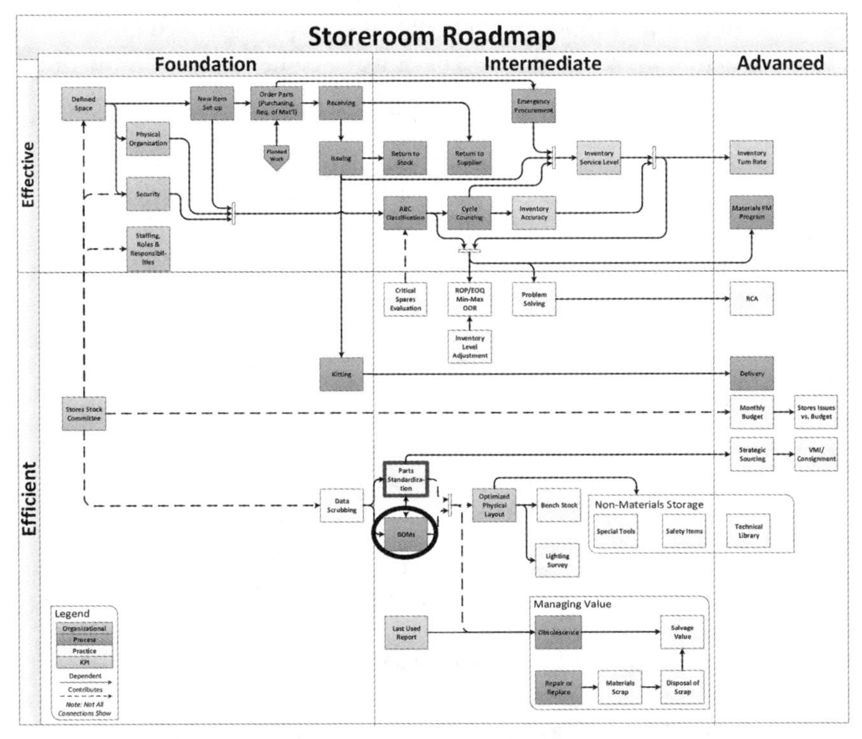

Figure 4-2 Bill of materials on the storeroom roadmap
Adapted with permission by Marshall Institute, Inc.

The position of BOMs on the storeroom roadmap shown in Figure 4-2 identifies it as a 'process' in the intermediate/efficient phase of a storeroom. Note that it is right below Parts Standardization, which was referenced in the last chapter.

As a 'process,' it is a tactical element, meaning that it is a topic that is almost a daily concern for the storeroom. That makes sense if you think about the complexity of 'what parts go to what machine?' The intermediate/efficient phase indicates that it is a process we would want to get better at once we have set some ground rules on exactly how to initially set items up in the storeroom and issue them out. New Item Set-Up and Issuing are foundation/effective phase processes and are located in the upper-left corner of the roadmap.

More directly, and to the point of BOMs and how they factor into the 'scheme of things,' is to consider how they impact, positively or negatively, almost every agency in our organization. Figure 4-3 is meant to illustrate just a small group of those who are affected by having a complete listing of the parts that make up an asset, or as we've come to know it, the Bill of Materials or BOM.

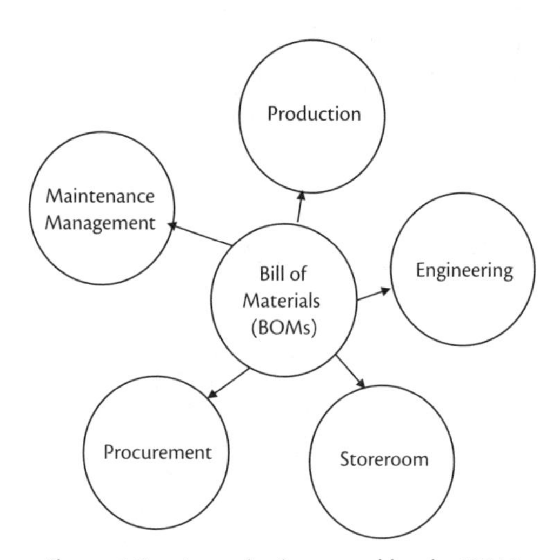

Figure 4-3 Agencies impacted by the BOM

I mentioned that Figure 4-3 illustrates just a few of the agencies that are affected by having an accurate BOM. Every group is affected by the BOMs, but we will stick with these five for additional study.

In the space provided, and in your own words, describe the interest that the named agency has in an accurate BOM. I'll start the first one for you so you can gain a sense of what is required for these entries.

Production: Production benefits from an accurate BOM for each asset through assurances that project engineering and corporate leadership have agreed to focus on parts standardization, presumably around the most robust parts. Additionally, through the world-class process of only stocking items for an asset that appear on that BOM, production is guaranteed that maintenance will only use the 'proper' parts.
Engineering:
Storeroom:
Procurement:
Maintenance Management:

As it turns out, and in the scheme of things, BOMs are the singular issue that links us all together around an asset. At a more granular level, BOMs link the engineering-heavy theory of *Equipment Design Excellence* and the maintenance-heavy theory of *Maintenance Excellence*.

Let's pause for a moment and reflect on some repeated *absolutes* mentioned in this book so far:

- Everyone is responsible for asset reliability.
- Everyone is linked to asset reliability.
- Properly running assets pay for everyone in the organization.
- Leadership is responsible for the organization having an asset management strategy.
- All agencies in an organization have an interest in the BOM.

Keep that in mind as we move on to discover what a BOM is.

BOM Not Bomb

A BOM is a listing of the parts that make up an asset. This idea, and the practice of this collection of 'what goes in it,' is alive and well in virtually every production facility on the planet.

Consider a factory that makes milk chocolate. It is a surefire bet that the company has the recipe written down for their signature chocolate. In fact, the associates involved in actually blending the chocolate are required to follow the recipe exactly. This has everything to do with making sure the brand's products taste right to the consumer, as it does with yield loss, cost overruns, and food safety. There is very little freelancing when it comes to the integrity of the actual products.

What about maintenance and reliability? Our products are technical advice and the top-notch service we provide daily. That service often requires that we have a spare part to replace a broken part. The presumption is that we know what parts 'came' with the machine to begin with, and we'll use those same parts in exchange. This goes back to the idea that assets arrive at our plant with an *inherent reliability*. It is the responsibility of everyone to guard that inherent reliability. One way we can do this is like-for-like component change outs.

We are likely to come across a BOM as an exploded-view diagram that is accompanied by several pages of call-outs for each specific item in the diagram. I usually reminisce with my classes about those old 1970s-era Chilton's Auto Repair Manuals when describing how a BOM might look in real life. Many of you reading this book might be familiar with those manuals. I'm certain that most of you have seen the exploded-view/call-out images I'm referring to. Let Figures 4-4 and 4-5 serve as reminders.

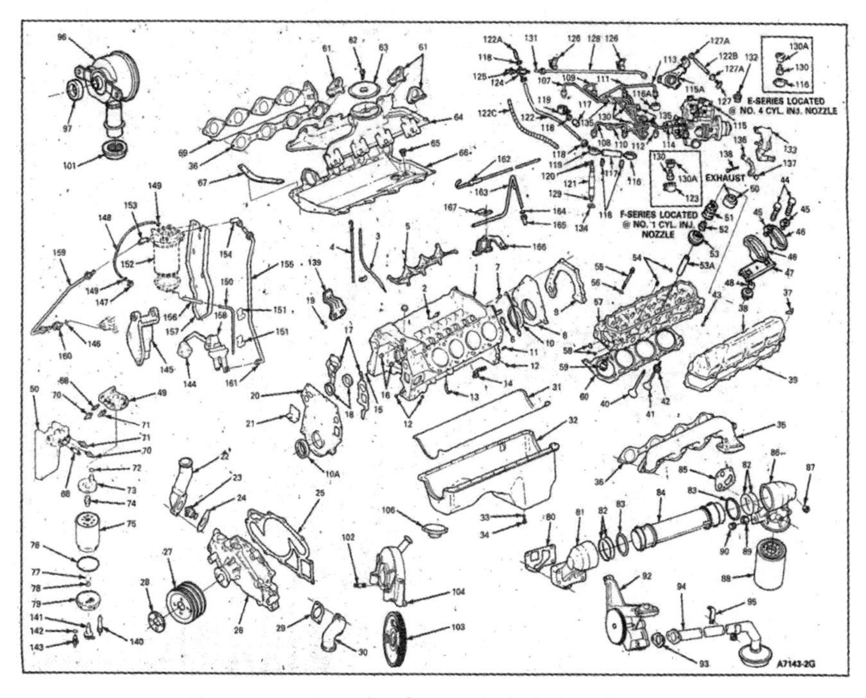

Figure 4-4 Sample of an exploded-view diagram

REF. NO.	BASIC PART NO.	DESCRIPTION
1	6009H	Cylinder Block Assy.
2	6C329A	Guide, Tappet
3	6754	Tube Assembly, Oil Level Gauge (F-Series)
4	6750	Oil Level Gauge (F-Series)
5	6C330A	Retainer, Tappet Guide
6	6026A	Plug, Engine (1-1/2" O.D.)
7	6D083A	Gasket, Rear Cover
8	6L060A	Cover Assembly, Engine, Rear
9	6A369A	Adapter, Flywheel to Transmission
10	6701A	Rear Oil Seal, Crankshaft
10A	—	Front Oil Seal
11	6B041B	Dowel Pin, Fly Wheel Adapter
12	87614S	Pipe Plug, 1/8 NPTF
13	6C327A	Piston Cooling Jet
14	6A051A	Heater Assembly, Block
15	6B041A	Dowel Pin, Front Cover Plate
16	6026E	Cup Plug
17	6020A	Gasket, Front Cover Plate
18	6A251A	Bearing Kit, Camshaft
19	6A628A	Ball, Oil Indicator Hole 11/32"
20	6B070A	Plate, Front Cover
21		Indicator, Timing (Part of Front Cover)
22	8592G	Connection, Water Outlet
23	8575	Thermostat
24	8255A	Gasket, Water Outlet
25	8507A	Gasket, Water Pump
26	8501D	Water Pump
27	8509D	Pulley, Water Pump
28	8546A	Spacer, Fan
29	8255A	Gasket, Water Inlet
30	8592D	Connection, Water Inlet
31	D6AZ-19562-A	RTV Sealant
32	6675C	Oil Pan
33	6734A	Gasket, Oil Pan Drain
34	6730A	Plug, Oil Pan Drain
35	9431B	Manifold, Exhaust, Left
36	9448A	Gasket Exhaust Manifold
37	6A532A	Washer, Valve Cover
38	6582C	Valve Cover
39	6584A	Gasket, Valve Cover
40	6507D	Valve, Intake (8)
41	6505	Valve, Exhaust
42	6057B	Insert, Exhaust Valve Seat
43	6026F	Plug, Ball Type 13/32" (8)
44	—	Bolt, Valve Lever and Washer

REF. NO.	BASIC PART NO.	DESCRIPTION
45	—	Post, Valve Lever
46	—	Lever, Valve
47	—	Retainer, Valve Lever Post
48	—	Lock, Valve Spring Retainer
49	6514A	Retainer, Valve Spring (36)
50	—	Shield, Oil (Exhaust)
51	6513B	Spring, Valve, with Damper (16)
52	6571A	Seal, Valve Stem-Intake (8)
53	6K533A	Rotator, Assembly Valve (16)
53A	—	Guide, Valve (Service)
54		Plug, 1/2 Inch NPTF (4)
55	6065A	Bolt Cylinder Head (34)
56	6L015A	Washer, Cylinder Head Bolt (34)
57	6049A	Cylinder Head Assembly (2)
58	6026B	Plug, 1/4 Inch
59	6057A	Insert, Combustion Chamber (8)
60	6051B	Gasket, Cylinder Head (2)
61		Eye, Lifting (3)
62	9C629A	Insert, Bolt Thread-Air Cleaner Stud
63	9F460A	Screen, Intake Manifold
64	9424B	Manifold, Intake
65	9A450A	Drain Plug, Valley Pan
66	9439B	Gasket and Valley Pan
67	9B470A	Strap, Valley Pan
68		Fuel Priming Valve and Cap
69	9430A	Manifold, Exhaust, Right
70		Continuous Vent with Check Valve
71		Vacuum Switch (Fuel Filter Element Replacement Indicator)
72		Fuel Heater O-Ring
73		Fuel Heater
74		Threaded Insert
75		Fuel Filter Element
76		Drain Bowl O-Ring
77		Drain Valve Stem Cap
78		Drain Valve Seal
79		Water Separator Drain Bowl
80	6A636A	Gasket, Oil Cooler, Front Header
81		Header, Oil Cooler, Front
82	6K649A	O-Ring, Oil Cooler (2)
83	6C610A	O-Ring, Oil Cooler (2)
84	6A642A	Cooler, Oil
85	6A636B	Gasket, Oil Cooler, Rear Header
86	6881B	Header, Oil Cooler, Rear
87	6K862A	Plug, 1/4-Inch
88	6731A	Oil Filter

REF. NO.	BASIC PART NO.	DESCRIPTION
89		Plug, 1/2 Inch
90		Plug, 1/2 Inch
92	6600	Oil Pump Assy.
93	6626	Gasket, Oil Pick-Up
94	6622	Pick-Up Tube
95	6A661	Bracket, Oil Pick-Up
96	6A665	CDR Valve
97	6A892	Seal Ring, CDR Valve
100	6758	Crankcase Vent Tube
101	6769	Grommet, Valley Cover
102	9F733	Mounting Stud, Injection Pump
103	9A546	Drive Gear, Injection Pump
104	9C516	Adapter Housing, Injection Pump
106	6766	Cap, Oil Filter
107	9A555H	Pipe w/Nuts Pump to Cyl. 8
108	9A555G	Pipe w/Nuts Pump to Cyl. 7
109	9A555F	Pipe w/Nuts Pump to Cyl. 6
110	9A555E	Pipe w/Nuts Pump to Cyl. 5
111	9A555D	Pipe w/Nuts Pump to Cyl. 4
112	9A555C	Pipe w/Nuts Pump to Cyl. 3
113	9A555B	Pipe w/Nuts Pump to Cyl. 2
114	9A555A	Pipe w/Nuts Pump to Cyl. 1
115	9A543	Injection Pump
115A	—	Valve, Vacuum Modulator (Auto. Trans.)
116	—	Fuel Return Tee
116A	—	Elbow, Fuel Return (F-Series)
117	—	Hose
118	9B255	Clip
119	9A564	Fuel Return Tee
120	87032-S92	O-Rings
121	9E527	Injection Nozzle Holder
122	—	Fuel Return Hose
122A	—	Hose
122B	—	Hose, Pump to Fuel Return Tube
122C	—	Guard, Rear Fuel Return Hose
123	—	Fuel Return Tee (E-Series)
124	9F734	Fuel Return Junction Fitting
125	—	Nipple, Fuel Return
126	9N659	Clamp
127	9F736	Elbow
127A	—	Clip
128	9D308	Tube
129	—	Nozzle Tip
130	—	Sensor, Fuel Line Pressure
130A	—	Cover
131	9C387	Sleeve Seal, Fuel Return (2)

REF. NO.	BASIC PART NO.	DESCRIPTION
132	9E939	Temperature Switch
133	12B526	Bracket and Solenoid-Fast Idle
134	—	Gasket Nozzle (8)
135	9N653	Clamp
136	9F541	Kickdown Lever (Auto. Trans.)
137	9D927	Screw, Kickdown Lever
138	9F539	Adjusting Screw (Kickdown Lever)
139	6786	Bracket, Oil Level Tube Support (F-Series)
140		Vent/Valve Assembly
141		Manual Drain Valve
142		Water Sensor O-Ring
143		Water Sensor Probe
144		Fuel Pump Supply Gasket
145		Alternator Bracket
146		Sealing O-Ring
147		Fuel Return Tee (At Nozzle)
148		Hose, 3/16" ID x 10" Long
149		Hose Clip
150		Water Drain Tube
151		Drain Tube Clamp (Z)
152		Fuel Filter/Water Separator Element
153		Elbow
154		Elbow, Fuel Supply Pump to Filter Header
155		Fuel Pump to Fuel Header Tube (With Two Nuts and Two Sleeves)
156		Hose, 3/16" x 2-5/16" Long
157		Fuel Filter Header Mounting Bracket
158		Fuel Supply Pump
159		Filter to Injection Pump Tube (With Two Nuts and Two Sleeves)
160		Connector Fitting
161		Inverted Flare Tube Nut
162		Oil Level Gauge — E-Series
163		Tube Assembly, Oil Level Gauge — E-Series
164		O-Ring, Oil Level Gauge — E-Series
165		Oil Level Gauge Tube, Lower — E-Series
166		Bracket, Oil Level Gauge Tube — E-Series
167		Retainer, Oil Level Gauge Tube — E-Series

Figure 4-5 Sample 'call-out' sheet in reference to exploded-view diagram

As an example of a BOM and how the value of such a resource might be of benefit to the entire organization, let's consider a simple asset that we all have some experience with, a simple ink pen.

Figure 4-6 is an exploded-view diagram of an ink pen.

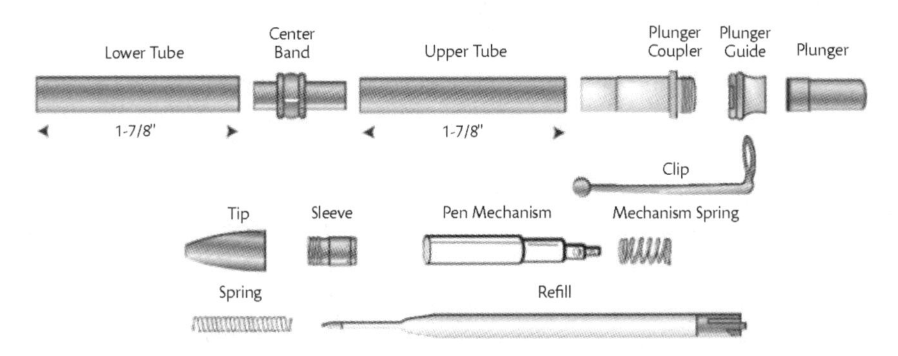

Figure 4-6 Exploded-view diagram of an ink pen

The simple example shown in Figure 4-6 will be used for various aspects of understanding what a BOM is and how BOMs should be used to assist with maintenance and reliability decisions. The ISO 55000 series of documents instructs us to have decision-making criteria and that we utilize data collection techniques and data management. The BOM delivers all that and more.

I actually do recognize that the pen shown in Figure 4-6 would need to be broken down a little further. The 'refill' depicted in the lower-right corner is made up of four sub-components: nib, tube, cap, and ink. Those are not shown as individual components in Figure 4-6, but in an exploded-view diagram we would pay for, I would expect to see these items detailed out. There are a few other smaller components that are hidden in Figure 4-6 as well. Can you find them?

The BOM for this simple pen is a listing of all the parts that are required to make the pen. Said another way, the BOM is a listing of all the parts (components) needed to build the asset (pen). I see eighteen individual pieces (components, parts) to this pen. See how many you can list (some are easy, some take a little imagination). Record your thoughts in Table 4-1.

Table 4-1 Your BOM Listing

1.	10.
2.	11.
3.	12.
4.	13.
5.	14.
6.	15.
7.	16.
8.	17.
9.	18.

Table 4-2 is the list that I came up with for Figure 4-6:

Table 4-2 My BOM Listing

1. Lower tube	10. Mechanism spring
2. Center barrel-tube	11. Pen mechanism
3. Center barrel-fitted sleeve	12. Sleeve
4. Upper tube	13. Tip
5. Plunger coupler-tube	14. Spring
6. Plunger coupler-fitted sleeve	15. Nib
7. Plunger guide	16. Ink tube
8. Plunger	17. Ink tube cap
9. Clip	18. Ink

Did you get eighteen? The two fitted sleeves are individual components.

A BOM might appear as a hierarchy diagram. Try your hand at constructing a parts hierarchy in the space provided in Figure 4-7.

Figure 4-7 Your parts hierarchy

Figure 4-8 is meant to represent my interpretation of Figure 4-6 as a parts hierarchy.

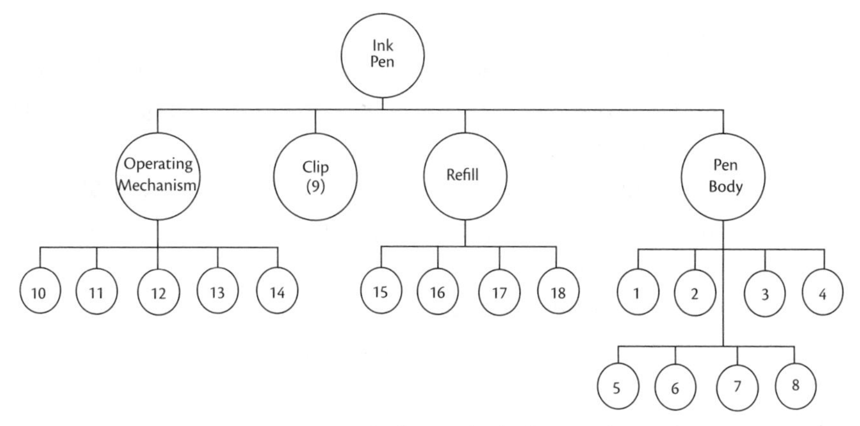

Figure 4-8 My parts hierarchy (refer to Table 4-2)

It's very likely that we've been exposed to a BOM as either an exploded-view diagram or a parts hierarchy. Each method is useful, and each gives us some sense and sensibility around what parts actually go into an asset. Better yet, the original BOM gives us the knowledge we need to determine what parts actually came on the machine to begin with. But who is responsible for providing this BOM?

Who Dat?

This is the worst kept secret of (almost) all time. The person responsible for bringing the original BOM into the company is the project engineer. Not just any ol' BOM, no, no. It must be the as-built BOM, provided in English, in writing and electronically as well.

I was going to build a pause into my text for the eye rolls I was certain would be coming with that last passage. Let's break this responsibility down a bit further.

Who would have more insight, knowledge, and control over the parts that are to be assembled to make up the new (or modified) asset: the project engineer, the maintenance manager, the planner, or the storeroom clerk? The engineer should have been the obvious answer.

The root cause of not actually receiving a decent BOM with a new asset or a modified asset is the originator of the project itself, and that is clearly on the shoulders of the project engineer. Let's prime this particular discussion with some direction from ISO 55002.

Asset management requires that the organization understand how asset-related decisions are made. This is critical and uber-important. The 'risk-attitude' of the organization is influenced by all factors, including internal and external forces. Solid decision-making criteria are of paramount importance in the infantile stages of a project to build a new asset, or modify an existing one.

But where are we really? There has been such abdication of the leverage of most companies, that any 'required components list' that we would submit to the winning OEM comes across as a 'suggested parts list.' I remember very distinctly a meeting I had with a corporate engineering manager for a Fortune 100 company. In an attempt to stress the point I was just making with you, the reader, I told the engineering manager that he should hand a lengthy list of parts to an OEM representative and say, "These are the parts you are to use on our new machine, and these are the parts that we'll pay to have on the machine." You're a Fortune 100 company for crying out loud!

If we haven't abdicated this completely, it's likely that we're willing to accept a less than sterling BOM from the OEM. In fact, the belief that OEMs want to be your parts provider into the future is more apt to be true than not. Refer to any BOM you might have received from the Original Equipment Manufacturer and it is probably vague and missing details on part specifications.

That makes sense if you think about it. Imagine asking an OEM to give you detailed, as-built drawings, and detailed as-built BOMs on a new machine. I think the OEM is correct in fearing that the next time you need a machine like the one you just bought, you may 'shop' their prints around.

I want to say this again: the project engineer has a responsibility to bring in, with the asset, a complete as-built BOM listing, provided in English, in writing and electronically as well. Once that document has landed in the plant, the maintenance planner is the person responsible for the continued upkeep and relevance of the BOM.

Just a quick closing note on this particular section to ask a few questions.

Would it be more advantageous to have a parts standardization process guaranteeing that we only accept assets with the kinds of components that seem to last around here?	Y or N
Is it logical to have a complete, detailed listing of the parts that make up the new or modified assets?	Y or N
Are we likely to refine and assemble really detailed information on components, including their failure histories and nuances, if we insist on standardization of parts at the very beginning of a capital project?	Y or N

I hope you answered yes to all three questions.

This is more of a rhetorical question. Shouldn't there be a connection between the BOM and what parts we stock?

Stock/Don't Stock

Let's get this straight, right out of the block. Speaking of only MRO spare parts, a storeroom should only stock components that 'go' to an existing asset in the plant. There are no, and should never be, exceptions. OK, maybe one. For example, as previously mentioned, there will always be a small percentage of items that have become obsolete for our purposes, through no fault of our own. That is the only exception. Those items remain in stock, with a code indicating that they are obsolete and not to be issued, until a time that they can economically and safely be disposed of.

If a part is found to be important enough, and 'could' be used, with some slight modification, to work on an existing asset, that part needs to be labeled as a suitable substitute for a particular cause and stocked as such.

With that major thought in mind, this statement should be at the forefront of your storeroom philosophy and center stage in your asset management policy: "We will only stock, as spares, those individual components that appear on an asset's Bill of Materials." Full stop, as the old Western Union telegraphs might say.

Here is a quick review. The project engineer brings in the BOM with the asset, or with a modified existing asset. Once the BOM is in the CMMS (Computerized Maintenance Management System), the maintenance planner is responsible for keeping the asset's BOM up to date. The storeroom personnel are not responsible for determining what to stock or not stock. However, they should and must reserve the right to cry foul when they suspect a part is being added to stock that doesn't go to an asset in the plant. I tell my storeroom classes to request firmly, "Show me where this part is on the machine."

I would feel comfortable in assuming that most people reading this book would want to know where the original round of suggested spare parts comes from. Our first encounter with a capital project is typically a list of suggested spares from the OEM. Unfortunately, like the abdication of our right to insist on the parts we will accept on a new or modified asset, we tend to 'roll over' on this as well. The OEM does not have your best interest in mind. They just don't. I'm sorry, but that is generally the case.

Instead, I encourage all organizations to develop a Stock/Don't Stock Decision Tree to determine if a component should be stocked as a spare part in the plant's inventory. I want to stress the requirement again from ISO 55002 that we use decision-making criteria and that we utilize data collection techniques and data management.

If you have experience with ISO standards, and as I mentioned in the opening section, simply document what you're going to do, do it, audit that you have done it, and have objective evidence to show that you have done it that way for some time.

The exercise for you now is to come up with a process that defines decision-making criteria and utilizes both data collection and data management techniques to drive your quest to stock a component or not stock a component.

I created a Stock/Don't Stock Decision Tree in early 2018 and it was published in a 2/2018 *MaintWorld* article accompanied by this note of wisdom on building an inventory using Reliability Centered Maintenance (RCM) fundamentals: "Just to recap, for the sake of this discussion on what to stock, or not stock, we will:

- Limit our MRO inventory value to 1% ERV
- Only consider items (for stock or non-stock) that appear on an asset's BOM
- First understand the asset's function and then the component's function" (p. 31)

Figure 4-9 (shown on the following page) is the decision tree I published, and it is my hope that you will develop and field a criteria and data-based thought process like this one to populate your MRO storeroom. The following text is an excerpt from *The Reliability Excellence Workbook: From Ideas to Action* and is meant to add further understanding to the decision tree.

> The very first section of the Stock/Don't Stock Decision Tree is concerned with our knowledge of the function of the component. Do we have any idea of what this part does and why it matters? That is basic Reliability Centered Maintenance.
>
> The second level of questioning is if the part exists on a BOM for an asset in the facility. If not, why are we even stocking it? It doesn't belong in a high-performing storeroom.
>
> If a part for consideration makes it past those two hurdles, do we know what function it serves on the asset it is assigned to? This helps us determine the degree of criticality we need to consider for this part.
>
> The real meat and potatoes of stocking an item or not comes in the next series of decisions. This is purposefully set up to really make us think of what the part does, and how we know where it is on the P-F curve. These are hard questions and need straight answers.
>
> Establishing a world-class storeroom and stocking it to perform as such is no pedestrian affair. This is serious work for serious people.
>
> We are limited as to what we can stock. If not limited by a cap restriction on inventory value, we're restricted by volume. For the storerooms that I've worked with, it is clear to me that we've used up all the square footage, and now we're working on the cubic footage. (pp. 320 and 322)

I'm especially interested in you, the reader, catching onto one of the first series of decisions to be made in this decision tree. Just to emphasize the ongoing discussion on BOMs, please consider Figure 4-10 as an expanded view of Figure 4-9.

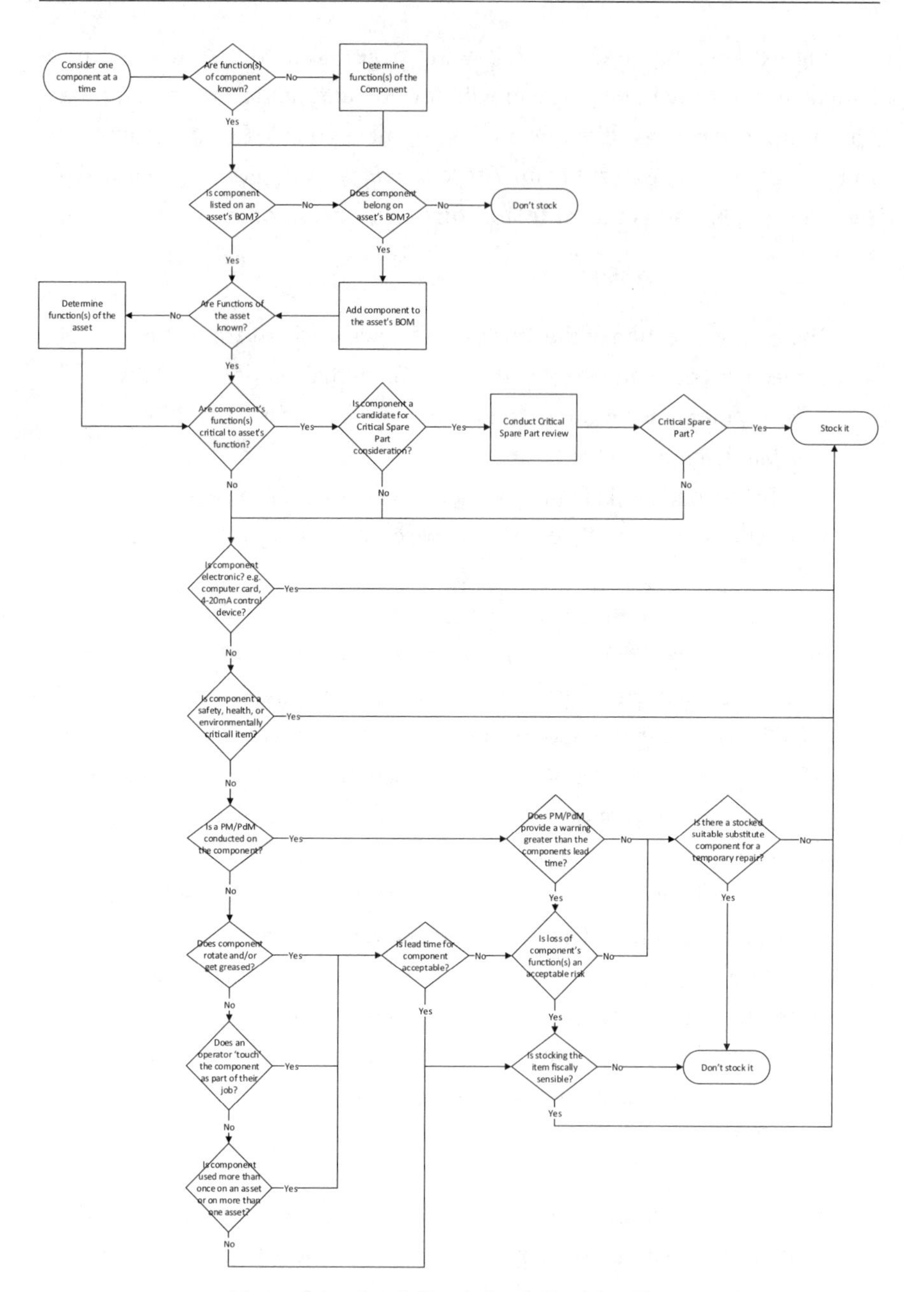

Figure 4-9 Stock/Don't Stock Decision Tree

Adapted with permission by Marshall Institute, Inc.

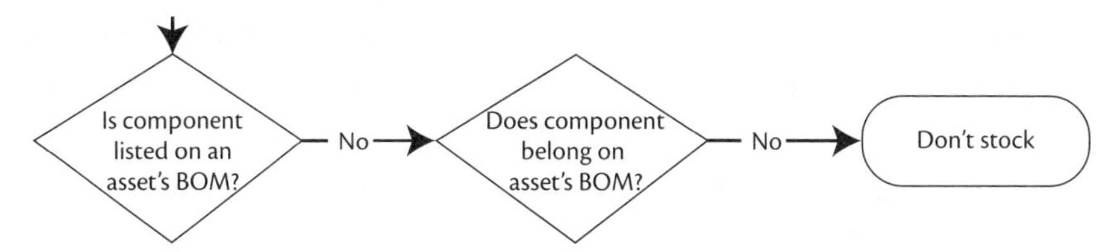

Figure 4-10 Emphasis on not stocking an item that is not on the BOM
Adapted with permission by Marshall Institute, Inc.

This point must certainly be 'driven home' by this time. Just don't do it. You will thank me later, trust me.

Now that we know what a BOM is, who is responsible for this listing, and how to use the BOM to determine what to stock or not stock in support of our asset's longevity, we should discuss methods for stocking items in our storerooms.

Different Strokes for Different Folks

I'm going to come right out and say something that might seem obviously wrong. If you disagree with me enough to want to write and tell me how much you disagree, contact my publisher. I promise I'll respond. Here goes: "If you're still stocking the majority of your components as minimum/maximum, you need to stop."

A min/max method just signals to others that we have no idea how many of one particular item we're likely to use. We have essentially given up on ever finding out.

Years ago, I was working with a midwestern company to take its storeroom from a very poor status-quo to a much higher level of service and convenience. When we got around to the stocking level discussion it was made clear to me that they stock components based on how many of that component they have in use in the plant. They literally were going for a 1:1 ratio. I rarely encounter a maintenance department that truly believes that they need to stock everything, but this was one in the making. I had a sense of *McHale's Navy*, if anyone recalls that old Ernest Borgnine television show from the 1960s. I would say that their minimum and maximum stock values were the same number.

Here's a little secret that very few people actually know. In fact it is likely that the storeroom knows this but has never communicated it to the maintenance department. When a maintenance organization tells the storeroom to stock an item with a min/max of a certain value, their input should have significant influence, but the real deciding source is the vendor or supplier. They determine what quantities they are going to sell the component in. The vendor will not likely break up a set or a pack.

In the space provided, list as many methods to stock an item that you can think of. We have already listed min/max; what else can you think of?

1.
2.
3.
4.

Here is my list for our discussion:

- OOR (Order on Request)
- EOQ (Economic Order Quantity)
- VMI/consignment (Vendor Managed Inventory)
- Safety stock (as a supplement to EOQ and min/max)

Order on Request

Order on request (OOR) is a very common and often appropriate 'stocking' method. The misnomer is that the item isn't actually stocked, but rather ordered in advance of when it is needed. This is a very powerful concept and very useful if used correctly.

Imagine that your engineers and your company are really adept at ensuring the BOMs are in place before the assets are ever put into production. From that BOM, and by utilizing a decision tree of what to stock or not stock, similar to the one shown in Figure 4-9, you have in place a pretty solid level of inventory in support of your plant assets. Clearly there are parts now remaining on the BOM that were not stocked, but will someday become necessary to order. These become the very items that we can set up as OOR in the inventory management system portion of our CMMS.

When an item is set up as an OOR item it is necessary to capture all the same details as if the item were being set up as a stocked item. All the information that you're required to fill in for the item master data for a stocked item is needed because you are literally going to set it up in the same database, except as an OOR item. Various CMMSs have their particular manner in which to accomplish this task.

Some clients actually carve out a tiny spot on their storeroom shelves, with the proper labeling to indicate, "This is where the part would be if we actually stocked the part." That seems odd, I'll admit, but it's a quick check for the storeroom and the maintenance planner to confirm that they actually *know* about the part and had *anticipated* its use.

The OOR process is not for the weak willed. If you can't keep your item master information straight on the parts that you actually stock, how good are you likely to be on items that you 'pretend' to stock?

If the ink pen depicted in Figure 4-6 were an operating asset in your plant, and its primary functions were to extend, write, and retract, which items would you stock, and of those that didn't make the grade, which of those would you set up on OOR? Record your thoughts in the space provided (the space provided is general and not meant to indicate that all the lines should be filled in).

Stock
Stock
Stock
OOR
OOR

With the functions listed, I would be inclined to stock the refill and have the two springs set up as OOR. I would run my Stock/Don't Stock Decision Tree and confirm my part's lead time, and also bounce that against how effective I would be at detecting a spring 'issue' in advance of any lead time. I wouldn't set up the clip as a stocked item or an OOR item because that isn't a function of the pen in my plant.

That last sentence is an interesting point if you consider that the exact same ink pen in another facility might have the following functions: clip to the pocket, extend, write, and retract. With those functions we'd have to consider the clip as a stocked or at least an OOR item.

So, what's the major takeaway from the discussion on OOR? Use a decision tree to stock or not stock an item. Of those items that are not stocked, determine the ones that are anticipated to be needed sometime in the future. Set those items up in the CMMS inventory management system as OOR items. You will need all the same information as if you were going to stock the items. The min/max will be 0/0. I recommend that you also have a solid sense of the lead time needed to secure the items.

Keep this point in mind as well: we will not stock parts in support of an asset that do not appear on that asset's BOM. We definitely will not set up items as OOR in support of an asset that do not appear on that asset's BOM.

I want to tie the OOR idea back to the ISO 55000 documents and make this argument. As it relates to the Strategic Asset Management Plan, it stands to reason that the *plan* would include access to the spare parts needed throughout the life of the asset, at least the asset's life while in our custody. Those parts must come from the BOM for the asset, and we understand that through the ebbs and flows of technology and obsolescence, the BOM will be modified by the planner, as needed. A Stock/Don't Stock Decision Tree, or some other manner to unbiasedly arrive at the best stocking solution, will be utilized.

This is accomplished in the arena of MRO spare parts to comply with the standard's requirement for decision-making criteria and utilization of data collection techniques and data management. If we are really good at decision making and creating information from the data collected, we might be successful with the next method of stocking.

Economic Order Quantity

Economic order quantity (EOQ) has as much to do with stocking the storeroom shelves as it has to do with purchasing the items to begin with. Or, to be more direct, the purchasing power that we can benefit from. This approach to storeroom inventory is familiar to anyone who has shopped at a 'big-box' store.

I'll share a secret with you about myself. I once was on a kick to save money, not for survival but to just see how much control I actually had over my own budget and finances. For an entire quarter (three months), I shopped for value. Not quality, but value. The quality aspect is why it only lasted for three months; it was not sustainable.

During this domestic adventure, I studied shelf labeling for everything I bought. I was determined to get the most product for the least cost. I realized shortly into this experiment that I may have gone off the deep end when I spent forty-five minutes in the toilet paper aisle, calculating which brand, roll size, and pillowy-softness gave me the lowest cost per individual *serving* of toilet paper. Note to readers: this is a product that you have permission to go high-end on. Do not skimp on quality.

What does toilet paper and stocking a storeroom have in common? That is a great question and deserving of a very good, yet concise, answer. Everything.

Controlling your storeroom inventory through an EOQ methodology has some significant mandates. First and foremost, only the components that have a historically predictable and stable usage pattern can be stocked through an EOQ calculation. Secondly, we are after the 'sweet spot' purchase price for quantity. This is the analogy I painted earlier about a big-box store experience. And finally, we have to figure into the equation the carrying costs of stocking the potential surplus of inventory, and fortunately we have a tool for that.

Back to toilet paper. Stable usage (check); high quantity for low cost (check); storage issues when stocking too much inventory (check and check).

There are many variations to the EOQ formula. I would recommend the calculation shown in Figure 4-11.

$$\sqrt{\frac{2NP}{IU}}$$

Figure 4-11 Common EOQ formula
Used with permission by Marshall Institute, Inc.

Whenever you see the square root symbol in a formula, that formula is important to know (rule of thumb). As you can see in Figure 4-11, there are five values at work:

1. 2 is a constant
2. N = annual usage of the item, full count
3. P = cost of a purchase order; most folks use an average of $65 per P.O.
4. I = carrying costs, percentage as a decimal
5. U = unit cost in dollars

The carrying costs is a puzzling number, as most organizations have no idea what their carrying costs really are. Reports that I've read seem to place an average carrying cost somewhere from 25% to 33%. That is a big range. You may recall that I previously showed you the results from a client's use of the potential savings calculator indicating their carrying costs. This is a very valuable number to know.

I mentioned earlier that by using the EOQ formula, we can determine the 'sweet spot' between buying in bulk and the cost of carrying all that extra inventory. Figure 4-12 graphically depicts this sweet spot.

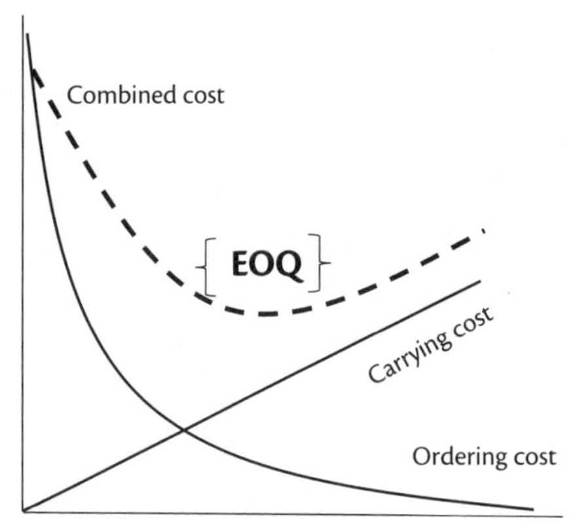

Figure 4-12 Graphic depiction of EOQ
Adapted with permission by Marshall Institute, Inc.

Consider the following scenario: Ace Manufacturing is a leading company in the southwestern United States. They have a lean operation and aggressively work to maintain a storeroom that is lean as well. A recent shift towards value-for-dollar purchasing has caused a ripple in the way things 'used to be.' A consultant on site recommended using an Economic Order Quantity approach. He showed them the formula in Figure 4-11.

The values they used to populate the EOQ formula are shown in Table 4-3 for a commonly used roller bearing.

Table 4-3 EOQ Exercise Data
Adapted with permission by Marshall Institute, Inc.

N	56
P	$65.00
I	0.25
U	$8.00

Use the formula shown in Figure 4-11 along with the data shown in Table 4-3 and calculate the ideal number of bearings to buy to keep the storeroom running lean. Record your answer here:

When I performed the calculation I arrived at 60.33 bearings, which I rounded down to 60. Ideally, we'd order 60 bearings at a time as a result of keeping carrying costs low, but also getting the 'most' out of a purchase order. A more real-life example that might be encountered in a scenario like this is with a vendor who only sells these bearings in quantities of twenty-five. It is common for a storeroom or purchasing group to come across this quandary. The formula tells you that you get the most value by purchasing 60 bearings at a time. However, the vendor tells you the quantity block they are actually going to sell you. I wonder how many folks reading this book would order 50 at a time and how many would order 75 at a time. How many would you order in this scenario? List that value here:

Min/Max and EOQ are only two ways to keep stocked items stocked. Another very common method engages the vendor to augment the storeroom effort.

VMI/Consignment

Perhaps the most familiar example of Vendor Managed Inventory (VMI) is the nut and bolt bins located just outside almost every storeroom. There is almost a 100% chance that your fasteners are managed and stocked by the supplier. This is VMI, where it is the responsibility of the vendor to clean, organize, and count the stock and create a stocking list from the shortages. VMI accounts are routine occurrences with the purchasing department and are usually set up on the standing purchase order protocol to lessen the red tape, and improve service and convenience (the two principle deliveries from VMI).

I had an introduction to VMI as a teenager growing up in Oklahoma. I worked at a grocery store during my high school years performing the usual duties such as sacking groceries, carrying bags to the customer's car (yes, that actually happened in my time), running price checks for the cashiers, and stocking the shelves.

During my tenure at the local market, it was common to be on shift when the soda distributors arrived to take care of the products on the shelves. If the Coca-Cola man showed up first, he'd rotate the stock, clean the remaining bottles, and stock the new additions behind the current stock. He never left without removing one row of Pepsi-Cola, sitting it on the floor, and replacing it with Coke. If the Pepsi-Cola man arrived first, he'd perform the same ritual, but he'd leave Coca-Cola laying on the floor. This went on for years. The Royal Crown man didn't stand a chance.

With VMI the *vendor manages* the *inventory*. Your company owns the inventory and you utilize the vendor's logistics and human resource assets. Nuts and bolts are the most common items to see set up as VMI, although few people recognize this as a vendor-managed stock. A benefit to having the vendor manage the inventory is that it reduces the number of purchase orders that are necessary, and it also reduces the hours of labor that storeroom personnel would be required to expend to straighten out and count inventory to make a replenishment list. This is all accomplished by the vendor.

A common mistake that most companies make is to think that only open-stocked items (fasteners, fittings, etc.) can be set up as VMI. Actually, any high moving item can be set up for vendor management. I always recommend that a storeroom oversight committee, which I call the Stores Stock Committee, be the approval body to set an item up for VMI. This should be a controlled type of stock.

More expensive and somewhat slower moving stock can benefit from a stocking technique called 'consignment,' which is the counter to VMI. With consignment stocking, the supplier still owns the stocked item(s), but the item(s) may be sitting in the manufacturing plant or facility storeroom.

The major benefit of consignment is believed to be financial. The company benefits by having a spare part instantly accessible on the shelf without having to pay for it up-front. The fiscal arrangement is a pay-and-play agreement. There is another major benefit as well—the immediate access to the part.

A major downfall to consignment is complacency. It would not be uncommon for a typical storeroom to be delinquent in the preventive maintenance activities that are required on stocked items (rotating motor shafts, greasing some fittings, ensuring electronic components are dry and cool, etc.). If a part on consignment is damaged, or found to be suffering abuse or poor care in a storeroom, the offending organization may have just then purchased that compromised part. You have to take care of consigned parts.

There is one consignment horror story I'll share with you. I have a business acquaintance who owns a surplus parts business. Years ago he was showing me a rather large >200 hp motor that he was holding (for a price) on consignment. A consigned part can also be stored on the supplier's shelf. Be very leery of this. I asked my friend who he was holding the motor for and he listed ten different companies. I questioned this, saying that I thought when things were on consignment for a company that meant only one company. My friend was being paid handsomely by ten companies that didn't know about each other for a motor that might actually not be available when needed. My friend was gambling and getting rich doing it. If your vendor is holding a component at their business, go and check on it from time to time.

One more stocking methodology that bears mentioning is safety stock.

Safety Stock

Let's get this straight. Your team has safety stock at all times. It's either formal and in the storeroom, or informal and laying at the bottom of someone's tool box. Safety stock is a legitimate concept for stocking an item that you feel at risk by not having. We all have those 'items' that we just add more to stock to ensure we don't run out of them. Running out of these kinds of parts could make an inconvenient situation even worse. Here's what that looks like.

> Suppler (of anything): "So you want one of these, right?"
>
> Customer: "Yes, but you better make it two, just in case."

Suffice it to say that items that we tend to hold a few extra of as safety stock are usually important and high moving items. The bearing example we calculated the EOQ value for earlier would be a great example. The EOQ formula netted an ideal ordering quantity of sixty. We might stock the sixty plus an extra five. There are formulas available that will lead you to an ideal safety stock and those formulas are meant to align with a world-class inventory service level value of 97%.

Safety stock is like the warning track on a baseball field. You don't want to run out of field at that point, but you do want to caution the player. For important components that move frequently, it is generally a good idea to have a few items formally on hand as safety stock. This type of stock must be limited and must be controlled. Figure 4-13 is a good example of how safety stock on some items can be of benefit to an organization.

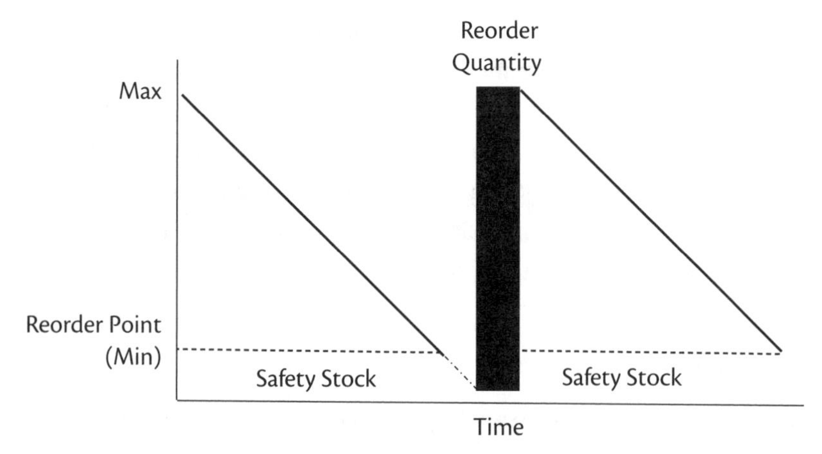

Figure 4-13 Safety stock

Note that in Figure 4-13, as the inventory is being depleted during the course of normal usage, the level of stock drops below the minimum. In this example, the minimum is zero and it is also the reorder point. While the organization waits for the replenishment, the safety stock is used to service any other issue requests.

A quick anecdote on safety stock. When my mom was a young newlywed in the early 1960s, she amassed a small fortune of $2,000. She told me once about this nest egg, which she kept secret, even from my dad. She had this huge (for the '60s) sum of money, hidden in plain sight in the family checkbook. She grew the fortune from the household allowance my dad provided from his military income. She deposited the windfall in the checking account but she never entered it in the account registry. When my mother passed away, she still had that same $2,000 from over fifty years earlier. That was the family's safety stock.

At this point in our discussion of BOMs and the storeroom, we now know that we only will stock items that appear on an asset's BOM. We've also talked about a Stock/Don't Stock Decision Tree. For those items we determine that we need to stock, we have several different ideas of how to actually stock them. Good stuff so far, but one question remains. Who is responsible for making sure this happens?

Whose Line Is It Anyway?

I quipped in my earlier book *The Reliability Excellence Workbook: From Ideas to Action* about the glaring issue we seem to have in almost all maintenance and reliability operations, and I'd like to expand that to include all asset management functions as well. That issue is the lack of clear delineation of roles and responsibilities. Here is what I wrote: "One primary detractor from good order and discipline in developing roles and responsibilities is the mere fact that we often think of roles and responsibilities as synonyms. They have very different meanings. The role, the specific role, actually aids in laying out the particular responsibilities of that role." (p. 149)

The storeroom's unique role is absolute asset management. I believe it can be argued that without access to the right part at the right time and in the right quantity, we (the leadership of an organization) are failing in the most primary tenet of asset management.

The Case

ISO 55002 states that organizations must assure that asset management has the same level of intensity and concern as other high-profile entities such as safety, quality, environmental, and other such socially and impactful objectives. In line with that, organizations are required to assess related risks against the assets and incorporate those in the company's risk management protocol. The mandate to ensure that everyone in the organization work to mitigate the risks associated with the continued operation of the assets in the facility is clear, and it starts at the top.

Keeping that in mind, let's also reflect on what should be the goal of the storeroom. I instruct my classes on maintenance and reliability practices that a storeroom is there to provide service and convenience. If the storeroom does not do both, you are better off not having a storeroom at all. In fact, I'd rather know definitively that the storeroom does not have a part than spend two hours looking for a part that isn't there. Examine the picture in Figure 4-14.

Figure 4-14 Empty storeroom shelves

What is wrong with the image in Figure 4-14 showing a storeroom with nothing in it? The short and correct answer is 'nothing.' If the storeroom is to provide service and convenience, how can an empty storeroom deliver on this promise? For all those instances that your maintenance department has no requirement from stores, the storeroom is delivering! When did we ever start to think that a storeroom had to have 'stuff' on the shelves in order to perform its required function?

What is the function of the storeroom, or what I like to call the goal of the storeroom? I mentioned it earlier: to have the right part at the right time and in the right quantity.

In this effort to have the correct material, when needed, there are many factors at play. The maintenance department has a responsibility to determine what to stock, what quantity to stock, and what to eliminate. I'll concede as a lifelong maintenance and reliability professional that we're likely to be very good at the first two elements and really lousy at the last one. For their efforts, the storeroom is responsible for keeping the requested items in stock and in the desired quantities. What dynamics are involved with setting up a stocked item in your storeroom?

New Item Set-Up

I've been working for years and throughout many industries to help set up or improve storerooms whenever and wherever needed. If you will refer back to Figure 4-2, in the upper left-hand corner, there is an activity block titled 'New Item Set-Up.' This is in the sector of Foundational-Effective. One of the principle activities to making a storeroom effective to start with is an ability to correctly set up new components in the stocked inventory. Figure 4-15 is a workflow template I use throughout all industries to get this right.

For this singular process, there are six agencies engaged in setting up a new item in the storeroom. Each of those agencies has a unique role to play. Remember my point at the beginning of this section that roles and responsibilities have two different meanings. Table 4-4 shows the six agencies and their individual roles.

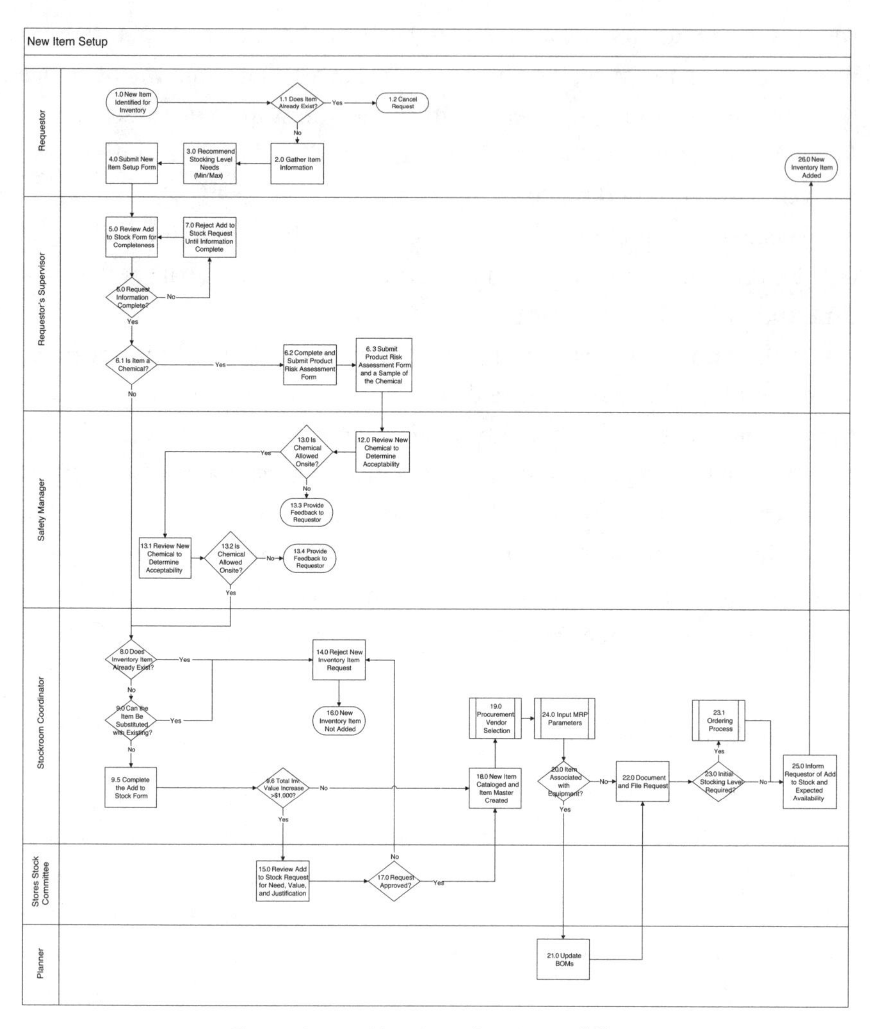

Figure 4-15 New Item Set-Up workflow
Adapted with permission by Marshall Institute, Inc.

Table 4-4 New Item Set-Up Roles

Agency	**Role**
Requestor	Need identifier
Requestor's Supervisor	Need evaluator
Safety Manager	Site chemical approver
Storeroom Coordinator	Inventory manager
Stores Stock Committee	Stock approver
Planner	BOM overseer

New Item Set-Up is such a crucial process for an effective and well-performing storeroom that I feel it's necessary to go into some detail. What follows is a series of specific responsibilities around each of these six agencies and their roles.

REQUESTOR

Figure 4-16 is the detail of the Requestor's responsibilities as per the workflow in Figure 4-15.

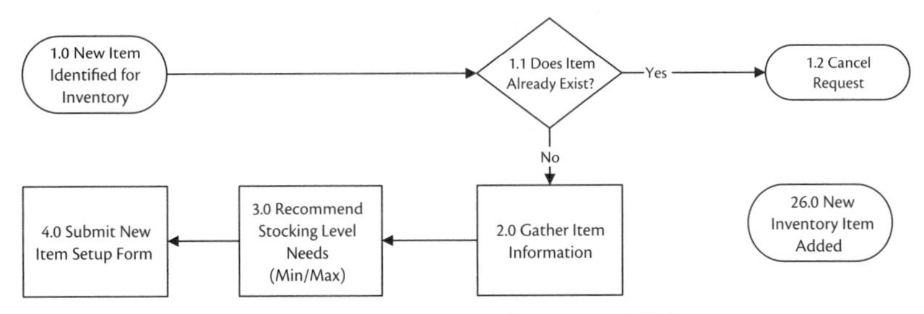

Figure 4-16 Requestor's responsibilities
Adapted with permission by Marshall Institute, Inc.

In a process guide, the responsibilities for any particular agency are reflected in the workflow, as in Figures 4-15 and 4-16, and are also depicted in the associated RACI chart.

RACI is an acronym that means: *Responsible*, *Accountable*, *Consult*, and *Inform*. The 'R' on a RACI chart indicates the person who has responsibility for that singular task. The 'A' is the accountable person, or typically that person's supervisor. Table 4-5 is the RACI extract for the Requestor in the New Item Set-Up process.

Table 4-5 Requestor's Responsibilities
Adapted with permission by Marshall Institute, Inc.

Index	Activity	Plant Manager	Stores Stock Committee	Maintenance Manager	Storeroom Manager	Stockroom Coordinator	Planner	Safety Manager	Requestor's Manager	Requestor's Supervisor	Requestor
1	New Item Identified for Inventory									A	R
1.1	Does Item Already Exist?									A	R
1.2	Cancel Request									A	R
2	Gather Item Information									A	R
3	Recommend Stocking Level Needs (Min/Max)									A	R
4	Submit Add to Stock Form									A	R
26	New Inventory Item Added									A	R

To summarize so far, in the New Item Set-Up process, the Requestor has the role of 'need identifier.' In that capacity, the Requestor has seven specific responsibilities as shown in both Figure 4-16 and Table 4-5. Each block in Figure 4-16 has activity requirements to complete within that individual block.

REQUESTOR'S SUPERVISOR

I'd like to follow the format you have just experienced by breaking down the different agencies involved in the remainder of the New Item Set-Up process and let you extrapolate the different responsibilities for the different roles. We continue with the Requestor's Supervisor shown in Figure 4-17 and Table 4-6. The Requestor's Supervisor has the role of 'need evaluator.'

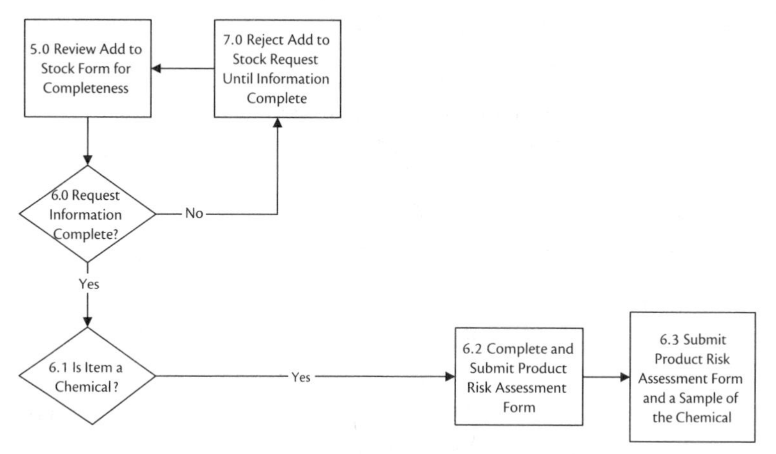

Figure 4-17 Requestor's Supervisor's responsibilities
Adapted with permission by Marshall Institute, Inc.

Table 4-6 Requestor's Supervisor's Responsibilities
Adapted with permission by Marshall Institute, Inc.

Index	Activity	Plant Manager	Stores Stock Committee	Maintenance Manager	Storeroom Manager	Stockroom Coordinator	Planner	Safety Manager	Requestor's Manager	Requestor's Supervisor	Requestor
5	Review Add to Stock Form for Completeness								A	R	
6	Request Information Complete?								A	R	
6.1	Is Item a Chemical?								A	R	
6.2	Complete Product Risk Assessment Form								A	R	
6.3	Submit Product Risk Assessment Form and a Sample of the Chemical								A	R	
7	Reject Add to Stock Request Until Information Complete								A	R	

SAFETY MANAGER

Just about every organization has a safety manager or at least a key person responsible for the administration of the plant or facility's safety program. For the New Item Set-Up process, the Safety Manager has the role of 'chemical approver.' Figure 4-18 and Table 4-7 show the detailed responsibilities.

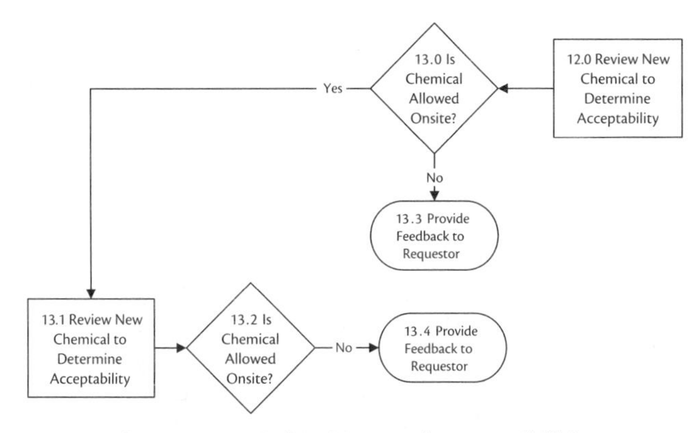

Figure 4-18 Safety Manager's responsibilities
Adapted with permission by Marshall Institute, Inc.

Table 4-7 Safety Manager's Responsibilities
Adapted with permission by Marshall Institute, Inc.

Index	Activity	Plant Manager	Stores Stock Committee	Maintenance Manager	Storeroom Manager	Stockroom Coordinator	Planner	Safety Manager	Requestor's Manager	Requestor's Supervisor	Requestor
12	Review New Chemical to Determine Acceptability	A						R			
13	Is Chemical Allowed Onsite?	A						R			
13.1	Review New Chemical to Determine Acceptability	A						R			
13.2	Is Chemical Allowed Onsite?	A						R			
13.3	Provide Feedback to Requestor	A						R			
13.4	Provide Feedback to Requestor	A						R			

STOREROOM COORDINATOR

Many different titles are often associated with this role. For the purpose of this specific work process, the Storeroom Coordinator is the person responsible for the activities around the new item being set up properly in inventory once it is confirmed that the part is indeed needed and necessary. The coordinator is also responsible for establishing the initial purchase for the item if required. See Figure 4-19 and Table 4-8.

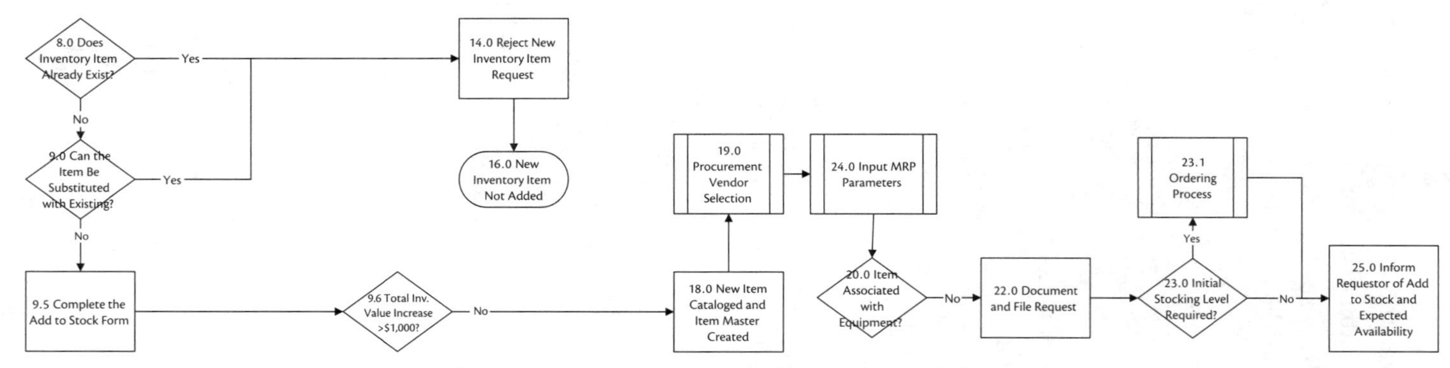

Figure 4-19 Storeroom Coordinator's responsibilities

Adapted with permission by Marshall Institute, Inc.

Table 4-8 Storeroom Coordinator's Responsibilities
Adapted with permission by Marshall Institute, Inc.

Index	Activity	Plant Manager	Stores Stock Committee	Maintenance Manager	Storeroom Manager	Stockroom Coordinator	Planner	Safety Manager	Requestor's Manager	Requestor's Supervisor	Requestor
8	Does Inventory Item Already Exist?				A	R					
9	Can the Item Be Substituted with Existing?				A	R					
9.5	Complete Add to Stock Form				A	R					
9.6	Total Inventory Value Increase >$1,000?				A	R					
14	Reject New Inventory Item Request				A	R					
16	New Inventory Item Not Added				A	R					
18	New Item Cataloged and Item Master Created				A	R					
20	Item Associated with Equipment?				A	R					
22	Document and File Request				A	R					
23	Initial Stocking Level Required?				A	R					
25	Inform Requestor of New Item Set-up and Expected Availability				A	R					

STORES STOCK COMMITTEE

The presence of a Stores Stock Committee (SSC) is the difference maker between an average spare parts program and a world-class program. It is the secret weapon. In this capacity, the SSC is the 'stock approver' of all New Item Set-Up requests. Figure 4-20 and Table 4-9 lay the SSC's responsibilities out for review.

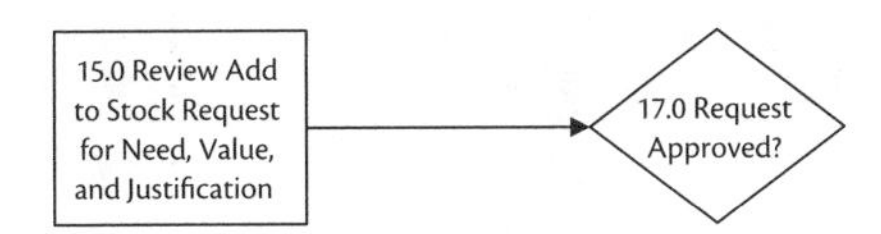

Figure 4-20 Stores Stock Committee's responsibilities
Adapted with permission by Marshall Institute, Inc.

Table 4-9 Stores Stock Committee's Responsibilities
Adapted with permission by Marshall Institute, Inc.

Index	Activity	Plant Manager	Stores Stock Committee	Maintenance Manager	Storeroom Manager	Stockroom Coordinator	Planner	Safety Manager	Requestor's Manager	Requestor's Supervisor	Requestor
15	Review Add to Stock Request for Need, Value, and Justification	A	R								
17	Request Approved?	A	R								

PLANNER

The Planner has one responsibility as the BOM overseer, and that is to simply update the BOMs. Please note that a component is never removed from the BOM. It may be hidden, but never remove an item from the BOM. The BOM becomes not only a list of the actual components that make up an asset, but also becomes a record of all the additions made over time. It is the history of the bits and pieces that make up an asset that creates the asset's timeline of parts usage at your location.

Figure 4-21 and Table 4-10 round out our detailed analysis of the New Item Set-Up process.

21.0 Update BOMs

Figure 4-21 Planner's responsibility
Adapted with permission by Marshall Institute, Inc.

Table 4-10 Planner's Responsibility
Adapted with permission by Marshall Institute, Inc.

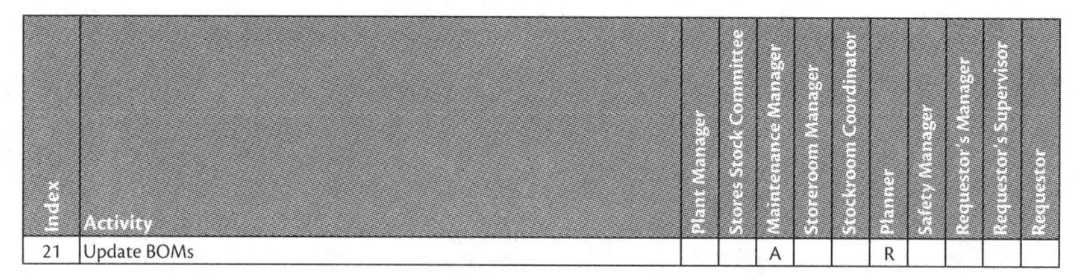

Index	Activity	Plant Manager	Stores Stock Committee	Maintenance Manager	Storeroom Manager	Stockroom Coordinator	Planner	Safety Manager	Requestor's Manager	Requestor's Supervisor	Requestor
21	Update BOMs			A			R				

Concluding with Figure 4-21 and Table 4-10, we have completed the detailed review of the six entities engaged in the New Item Set-Up process and listed out their responsibilities within their specific roles. This is just one of at least twenty-eight processes for an effective and efficient storeroom, which provides service and convenience by having the right part, at the right time, and in the right quantity.

The Not-So-Dirty 2-1/3 Dozen (28)

I tried to get cheeky with the title to this section but I see now that I missed the mark completely. I want to tie this chapter up with a summary of the storeroom's absolute importance to asset management for everything we need in our plants and facilities, in addition to complying with the ISO standards and for whatever an ISO 55000 certification would mean to your organization.

First you must accept the premise that access to spare parts is not keeping pace with advancements in technology. ISO 55000 and just common sense tells us that we have an obligation to shareholders and stakeholders alike to reduce our risk exposure to getting outpaced by a technology and a margin-call on our businesses to eke out more profit and reduce costs.

Including the New Item Set-Up process detailed earlier, there are twenty-eight processes that are required to keep the parts and components in our storeroom fresh and relevant:

1. ABC Classification Review
2. Adjust ABC Classification Model
3. New Item Set-Up
4. Bench Stock
5. Critical Spare Parts Algorithm
6. Critical Spares Review
7. Cycle Counting Criteria
8. Cycle Counting
9. Data Scrubbing
10. Disposal of Scrap
11. Document Stock Out
12. Emergency Procurement
13. Issuing
14. Item Substitution
15. Kitting
16. Obsolescence Criteria
17. Obsolescence
18. Optimizing Stock Levels
19. Ordering
20. Preservation Program
21. Receiving
22. Repair or Replace
23. Return to Stock
24. Return to Supplier

25. Salvage Value
26. Special Tools
27. Standardization
28. VMI/Consignment

Do you recall my warning at the very start of this chapter? "One necessary element, buried in ISO 55002 to be specific, is a section on management of change, with a small note that the organization should address supply chain constraints as a means to avoid an increased leveraged risk by this particular choke point (that being the lack of access to spare parts). This is rationale enough to address at the highest level what most of us have known for a very long time. Namely, that spare parts, or the dearth of them, is our Achilles heel!" If you remember that warning, take heed now.

The next storeroom I visit in hopes of delivering that client's desire for a world-class storeroom is likely to have just one person in it. One storeroom clerk who, according to everyone, is victimizing the maintenance department. One soul against an army of the oft disgruntled trades. How can it be that one person, or even a small(ish) group is left to execute twenty-eight processes at the highest level? Great storerooms, good storerooms, average storerooms and lousy storerooms have one thing in common: twenty-eight processes.

In terms of asset management, no asset management plan or Strategic Asset Management Plan (SAMP) is complete without a rock solid understanding of the intricacies of a maintenance storeroom and BOM.

CHAPTER SUMMARY

Chapter 4 focused on the BOM. More than just a discussion on BOMs, we also gained a greater sense of how a storeroom and the processes of an effective storeroom contribute to the asset management plan. It probably came as no surprise to anyone that this is one of the most critical aspects of reliability, maintainability, and the SAMP. Virtually every aspect of an asset is within our control except for spare parts.

Think about this: we design equipment, and manage the construction, validation and start-up. We determine in our organizations how we are going to use the equipment, for how long, and with how much maintenance. We even determine when the equipment is at the end of its life. We might have a BOM, but we don't fabricate a single part on a machine that the machine needs in order to run. We control everything except access to the spare parts. This is crucial.

Adding to Our Business Case

We recognize that access to the correct spare parts, when needed and in the correct quantity, contributes to the overall health and reliability of the assets. Furthermore, since assets arrive at our facility with inherent reliability, it is the role of everyone to ensure that the inherent reliability is maintained. One way to accomplish this is to have a supply chain guaranteeing all the spare parts when needed.

The ISO 55000 series of documents is riddled with pronouncements that the organization must have systems and processes in place to measure, evaluate, and essentially deal with risks of all kinds. Who can deny that the all-consuming, time-intensive process of locating and procuring the necessary spare parts is putting our organization at risk?

One necessary element, buried in ISO 55002 to be specific, is a section on management of change, with a small note that the organization should address supply chain constraints as a means to avoid an increased leveraged risk by this particular choke point (that being the lack of access to spare parts). This is rationale enough to address, at the highest level, what most of us have known for a very long time. Namely, that spare parts or the dearth of them are our Achilles heel!

Using a Stock/Don't Stock methodology, we will determine what components to stock in our on-site storeroom. We will only stock items for an asset that appear on that asset's Bill of Materials, or BOM. A BOM is a list of all the individual components that make up the parent component. For each new or modified asset, we will demand an as-built BOM, provided in English, in writing and electronically. These BOMs will only be accepted for the 'as-built' asset.

A BOM is often depicted in an exploded-view diagram such as this:

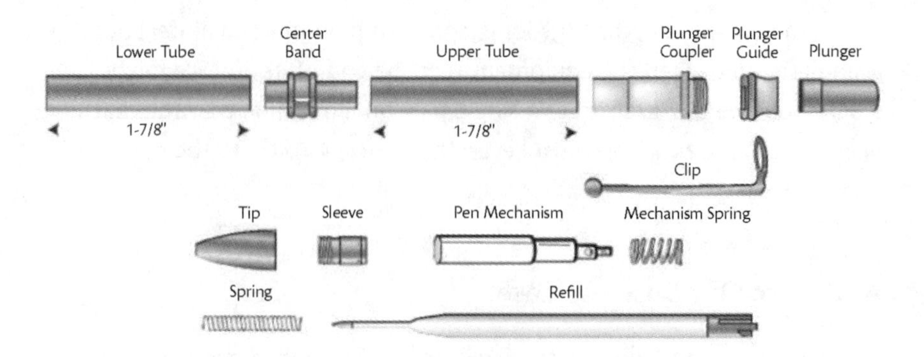

The diagram is accompanied by a listing of all the components on the asset, such as this:

1.	10.
2.	11.
3.	12.
4.	13.
5.	14.
6.	15.
7.	16.
8.	17.
9.	18.

Our organization will retain the as-built BOMs for each asset and the associated parts listing. Further, these parts will be arranged into a parts hierarchy similar to the following:

The project engineer assigned to the capital project to purchase a new asset, or modify an existing one, will have the primary responsibility to deliver the complete BOM thereby meeting the criteria recently defined. The planner has the responsibility to oversee the BOMs with updates and changes as the asset is utilized and maintained while in the performance of its function in our facility.

A proper and complete BOM has a positive and lasting effect on almost all plant agencies. Following is an example of the agencies that benefit from a BOM:

Production:
Engineering:
Storeroom:
Procurement:
Maintenance Management:

For those components that appear on an asset's BOM, and which we determine should be stocked in the storeroom, we will evaluate and optimize the stocking method. For sporadically used parts, the best method to use is the traditional min/max level. We will apply a concept known as safety stock where appropriate. Beyond min/max, we hope to identify those components that have a stable usage pattern. For those we will apply the Economic Order Quantity (EOQ) approach using the following formula:

$$\sqrt{\frac{2NP}{IU}}$$

- 2 is a constant
- N = annual usage of the item, full count
- P = cost of a purchase order; most folks use an average of $65 per P.O.
- I = carrying costs, percentage as a decimal
- U = unit cost in dollars

Where it makes sense in terms of preserving the asset and its functions for the value of our stakeholders and our shareholders, we will work with our vendors and suppliers to establish Vendor Managed Inventory and/or consignment stock.

In all our storeroom and spare parts efforts, we will formalize the twenty-eight world-class storeroom processes and resource our stockroom so the level of our MRO capability matches what is needed to meet the organizational objectives. In each process there will be clearly drawn roles and associated responsibilities.

Turn to the last section of the book and record this same information in the comprehensive strategy section.

FIVE

Production in an Asset Management World

"Make sure every job has somebody to do it."
—Fergus O'Connell, *Simply Brilliant* (p. 59)

Doesn't it make sense that such a short and to the point quote as the one heading this chapter would come from a book titled *Simply Brilliant*? For this chapter on the production department's roles and responsibilities in a world governed by a formal asset management plan, it is imperative that each new(ish) asset-based assignment given to the production department have a qualified, and better yet, interested person to perform the job. An idea was established at the start of this book that we all have a role to play in asset reliability and indeed asset management. I even gave the example of how we're all connected to the asset in a way similar to the six degrees of separation party game. We owe our very paychecks to the proper and reliable functioning of the company's assets. Would it make sense that the group that has the largest number of human assets in the plant or facility have a significant role to play in the care and upkeep of those very same assets? I think so and I bet you do as well.

This idea of an organization whose stakeholders squarely have 'skin in the game' transcends capital assets as evidenced in Alice Schroeder's book, *The Snowball*, a biography of investment icon Warren Buffett. Buffett and his Berkshire Hathaway partner, Charlie Munger, wrote in an annual shareholder's newsletter, "We do not view the company as the ultimate owner of our business assets, but, instead, view the company as a conduit through which our shareholders own the assets." (pp. 409–410)

We are in this together. Maintenance, production, everyone.

The challenge is to chart a new course in ownership within our production departments. This course will include a prominent role for the operators to play in an effort to meet the organizational objectives through smart and engaged associates. Where do we start? As Julie Andrews advised us in *The Sound of Music*, "Let's start at the very beginning, a very good place to start." (Rodgers, Rittmann, 1959)

Begin with the End in Mind

I'm pretty sure I borrowed this section's title from *The 7 Habits of Highly Effective People* (Covey), but it dovetails nicely into this discussion. Can we agree on the notion that a well-intended, but poorly trained operator can really screw up a piece of equipment? There must be recent evidence of this very occurrence at your location. Perhaps these events are a reflection that our operator training needs some enhancement. In general, a shortcoming in the production department's overall engagement in seeking out, finding, and reporting asset defects might be an indication that our operator onboarding process is lacking.

I want to ensure that the core of this book's purpose is kept out in front of us while we are creating our thoughts around an asset management philosophy. ISO 55002 instructs the organization to establish processes (including operational) to support the asset management plan. Those processes are to include roles and responsibilities, resources, and competency development. Even if your organization has zero interest in an ISO 55000 certification, you can't argue that the concept of educated production involvement in asset care isn't important.

A Desert of Knowledge

Here is an undeniable truth as it relates to the competency of our operators. Similar to the dearth of skill in the skilled trades, and the global concern that "we just can't find qualified technicians," we must also consider the desert that production finds themselves in. There is almost a 100% chance that the next operator applicant at your location has zero experience with your equipment. It is very likely that they have very limited industrial experience at all. Ask yourselves this rhetorical question, "Is our operator training the best training in the world?" I think the course to greater operator involvement begins with the onboarding process. Set the expectations high at the beginning of an associate's tenure, and don't lower the bar to meet the applicant's qualifications.

Get on Board

My daughter attended a small, private liberal-arts college in Missouri. This was an all-girls college and the size of the school and the curriculum were exactly what she needed to launch her professional life. During the orientation week of her first year, the freshmen girls arrived on campus a few days early to get oriented and situated before the seniors came in to begin what must have been the usual big-person/little-person hazing rituals.

At the beginning of orientation week, all the young ladies were ushered into the campus auditorium for the standard staff and council introductions. After a series of department heads gave their boilerplate greetings, a member of the campus maintenance department went to the podium and began a discussion with the freshmen girls on how to notify the maintenance department if there were any issues the girls might have with the facilities while attending school. He showed them how to fill out a work notification online using the school's website and their student ID numbers. Hearing that experience from my daughter really gave me a reason to question how we introduce associates to their responsibilities at our locations.

It seems to me nowadays that during the onboarding process an employee is likely to get more instruction on how to fill out a 'vacation request form' than they do on how to fill out a 'work request form.' That needs to change. Odds are that an associate will need to request work from maintenance more often than they will vacation days from their boss.

This must be a cautionary tale, however; the answer is never as simple as showing operators how to fill out a work notification or work request. An operator must *first* have a sense of the theory of operation of their equipment before they can *then* determine that something is wrong with it.

At your location, briefly describe how an operator is trained on the theory of operation for their machine:

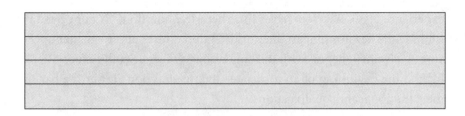

Read into this that we have a requirement to build competency in our operator corps. That competency starts with an absolute understanding of how the machine functions, what the components of the machine are, and the operator's roles and responsibilities around not only operating the asset (the primary role) but also ensuring the reliability of the asset (secondary role). This makes sense even outside of ISO 55000.

The ISO 55000 series is very consistent with this message:

- Establish organizational objectives.
- With the stakeholders (which includes employees), determine the asset management plan focused on achieving those objectives.
- Determine the competencies necessary to complete the roles and responsibilities.
- Conduct a gap analysis to see where the current competencies are wanting.
- Close the gap in line with organizational objectives and the Strategic Asset Management Plan (SAMP).

The onboarding process for production associates should include an unambiguous exchange of the new employee's responsibilities as they relate to the organizational objectives, their responsibilities toward the care and operation of their assigned asset(s), and a very comprehensive plan to measure and assess their knowledge and development of the training protocol to close the gap of their current understanding and company expectations. Understanding how the asset works is much more than understanding how to work the asset. This theory of operation is the central key to communication inside and outside the production department.

In this effort to develop competencies, production might be fighting a structure that is fortified against them.

Circle the Wagons

One of the very first magazine articles I ever wrote centered on how our traditional maintenance organizational structure supports our reactive tendencies. Bear with me while I try to make a similar connection with how a production department is likewise organized. I think I can make this work.

Here was my thesis as it concerned the way maintenance traditionally arranges its human assets: "Consider this too, we might in fact be passively contributing to the cause of our reactive nature without being aware. For example, I have noticed in American manufacturing that there is an overwhelming ignorance in establishing a proper maintenance structure. We have by default, structured ourselves to be wholly reactive." (Ross, *Uptime*, p. 32)

Give some thought to a stereotypical manufacturing plant that has followed an almost comical (by twenty-first century standards) 'breakdown' reporting protocol.

- Operator has a problem.
- Operator calls for maintenance on the radio.
- Operator waits for maintenance.

Rhetorically, does this seem reactive or proactive? It is simple and basic, but also true in many companies around the United States today. An approach that is starting to take a foothold in some manufacturing and service industries is the idea of work escalation. This is fundamentally the approach to move from reactive to proactive, but comes with this huge warning label:

'IF YOU'RE NOT GOING TO DO IT RIGHT, DON'T DO IT'

Let's bring in a practical example, and I'm fully prepared for many of you reading this book to prove me wrong. In fact, I hope that you do. In the space provided, document the steps that are executed to solve a small mechanical issue on a production asset at your plant. Let me set the scenario.

An operator notices a small hydraulic leak on his machine. The concern is that the hydraulic leak may contaminate the product. The fluid is leaking at a rate of one drop per minute. At your plant, how is this communicated and what are the steps taken to address this concern?

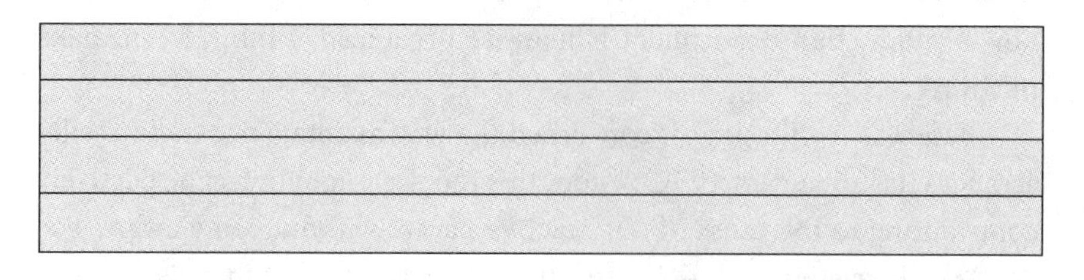

As I mentioned, I hope many of you prove me wrong on this idea that we are structured in production to be reactive. What follows is a typical process I have seen played out around the world, and then immediately following that example is another one where the organization has adopted a philosophy of maintenance escalation.

Traditional

An operator notices a small hydraulic leak on his or her machine. The concern is that the hydraulic leak may contaminate the product. The fluid is leaking at a rate of one drop per minute.

- The operator calls for maintenance to respond to the machine to address and hopefully stop the leak.
- Maintenance does not respond immediately, nor does maintenance notify the operator that, 'we heard you and we'll get to you soon.'
- The operator notifies his or her supervisor of the issue and confirms that they have already called maintenance to no avail; the supervisor may or may not enter a work request.
- The next day, the operator notices the same leak and puts in another call to maintenance and tells the supervisor one more time. The supervisor may or may not enter a work request (again).
- Day three, same leak, the operator calls it an emergency or safety issue to get the ball rolling.

- A maintenance technician is pulled off of a preventive or planned maintenance activity to go tighten a fitting.
- The operator, the supervisor, and the technician complain about each other.
- Situation normal.

Maintenance Escalation

An operator notices a small hydraulic leak on his machine. The concern is that the hydraulic leak may contaminate the product. The fluid is leaking at a rate of one drop per minute.

- From his training, the operator checks to make sure the hydraulic pressure is within the proper range and then uses a small wrench to tighten the fitting; the leak does not stop.
- The operator calls his supervisor (the first level of escalation).
- The supervisor confers with the operator to understand where the troubleshooting began and ended.
- The supervisor instructs the operator to shut down the asset, the system pressure is bled off, and the machine is properly locked out.
- From her training, the supervisor disconnects the hydraulic hose and inspects the fittings and the O-ring. Everything seems to be in good shape.
- The supervisor reassembles the hydraulic connection and instructs the operator to start the machine back up.
- The supervisor had hoped that merely reseating the fitting might have solved the issue; it did not.
- The supervisor creates a work order and calls maintenance (based on potential product contamination). *Please note that the supervisor created a work order and not a work request.
- The maintenance Do-It-Now (DIN) crew responds to address this situation.

Many organizations have converted to this more 'proactive' approach by adopting an escalation of maintenance procedure. Note that in each case, the production individuals are performing their duties as trained (operator then supervisor). Your organization would not benefit from unskilled and untrained operators or supervisors taking it upon themselves to fix equipment issues. I've given the example of the horrifying and hopefully unlikely sight of rounding a corner in a plant somewhere to find an operator on the floor disassembling a planetary gear box. That would not be a soothing image.

There is a popular trend among the more progressive companies to create and staff an operator-technician position. In fact, that role commonly goes by the moniker 'operator-tech.' The advent of this sort of forward thinking, in terms of asset reliability and capability, is directly on course with ISO 55000's strategic thinking, not to mention the morale boost it offers locally.

Consider that it is the facility's maintenance personnel who are often the most highly paid hourly employees and have the many benefits of flexibility and training opportunities. Can you conceive of the positive influence such roles of note would have within the operator group? This would offer another chance of advancement and relevance.

Through the addition of an operator-tech position, the maintenance escalation process would now progress through several people in production before a maintenance person was diverted from other planned and scheduled work. The progression might look like this:

- From the Operator to his or her Supervisor
- From the Supervisor to the Operator-Tech
- From the Operator-Tech to the Maintenance Technician

Would this suggested approach to an organizational structure of the production department increase, decrease, or keep the same the operational engagement in asset reliability? Would such a formal philosophy be likely to increase, decrease, or keep the same the level of the organization's commitment to theory of operation training and operator's skills development?

Did you answer 'increase' to both questions?

Training and competency seem to be where common sense meets ISO 55000 for reliability and asset management.

Welcome to the Matrix

In many ways our progressive companies have failed to progress. Not in all areas, of course, but in the investment in our human capital. Really think about this point for just a moment and ask yourselves if we have really engaged our associates completely and have made them true 'stakeholders,' to use a term from ISO 55000. Recall that all our employees are going to be involved in aligning organizational asset management objectives with the activities needed to get the job done. I think this quote from Deming in *Out of the Crisis* might describe where we find ourselves: "The greatest waste in America is failure to use the abilities of people." (p. 53)

Recall this point made earlier in Chapter 1: ISO 55000 directs those adhering to this standard that an organization's top management, *employees* and stakeholders are the groups responsible for conceiving and executing what is referred to as "control activities." These activities might include policies, procedures, and performance measuring and *monitoring techniques*. Would you agree that the best candidate to 'monitor' an asset day to day is the operator?

Figure 5-1 (on the following page) shows the increase in job responsibilities generally associated with operators over the last seven decades as a result of popular movements. What traditionally started as an "I operate it, you fix it," approach has matured to a level where production associates are assuming more and more upkeep responsibilities under the guise of autonomous care and beyond.

I think what is truly fascinating about Figure 5-1 and what it implies is that we didn't think to include operators in reliability and maintenance activities to begin with. Our first educated opportunity was when Total Productive Maintenance came ashore in the 1980s. Unfortunately, American industries keyed in on the word 'Maintenance' and quickly made TPM a maintenance program, and not the production-focused program as might have been intended by the author of the concept, Seiichi Nakajima. Regardless, here we are at a time in history where our bench strength in both maintenance and production is so light that we need "all hands on deck." Our depth of qualified associates is of such a veneer thickness that we have inadvertently put ourselves at a great risk.

Operator Engagement	Industrial Complex—blue color vs. white color			TPM Introduced	TPM Lands in the USA	Lean Manufacturing	Downsizing	Autonomous Care	ISO 55000
ERA	**1940s**	**1950s**	**1960s**	**1970s**	**1980s**	**1990s**	**2000s**	**2010s**	**2020s**
World Events	WWII Repatriation of Veterans	Korean War Strong Unions Cold War	Vietnam War Anti-Establishment Civil Rights Space Race	Oil Embargo	End of Cold War	Gulf War Dot Coms	Recession Y2K 9/11	War in the Middle East	

Figure 5-1 Operator engagement over time

How do we improve from here, and more than that, how do we strategically set our future selves up for success when we ourselves are long gone or at least have moved on to other opportunities?

The good news is that we have a blueprint for how to increase the knowledge and capabilities of our operations corps and we have a mandate to do so with the ISO 55000 standards. Let's unpack a Chapter 1 refresher from a few paragraphs earlier. I am going to edit it down to just the passage we are focusing on in this chapter.

ISO 55000 directs those adhering to this standard that *employees* are the group responsible for conceiving and executing what is referred to as "control activities." These activities might include: policies, procedures, and performance measuring and monitoring techniques.

It is now incumbent on 'us' the leaders, to make sure our employees (read 'employees' as 'operators') have the mental wherewithal to understand their equipment, become qualified enough to participate in the activities necessary to ensure proper operation and reliability, and to conduct the measuring and monitoring activities to make sure their assigned assets are functioning for the good of the organizational objectives.

This concept is unique to every organization. Compounding the issue is the myriad of organizational structures, capabilities, and allowances (union and non-union companies) present within each company and at each location. ISO 55000 refers to this as *operating context*.

This section is titled "Welcome to the Matrix" to emphasize that a training matrix is the method that will be used to comply with this specific need. Many of you reading this book are likely familiar with the training matrix format introduced in Chapter 1. Later in this section I'll provide you with an example that you might find helpful for operators.

Here is a short exercise that will align your thinking to what is required for this particular objective. Think about the title headings that follow and what activity under each heading would be beneficial for an operator to be capable of executing. I will provide some examples to 'prime the pump' so to speak, so fill in your answers right behind mine.

1. What asset-related policies might a production employee be engaged in the conception of at your facility?
   Example: Operators will be required to determine the appropriate dress code while attending to operating equipment, based on company safety protocol.

2. What asset-related policies might a production employee be responsible for executing at your location?
   Example: Production employees will complete asset run logs prior to shift change.

3. What asset-related procedures might a production employee be engaged in the conception of at your facility?
   Example: Production employees will create the cleaning, lubrication, and operator inspection procedures for their assigned asset.

4. What asset-related procedures might a production employee be responsible for executing at your location?
   Example: Operators will execute line-clearing procedures before all asset start-ups.

5. What asset-related performance measuring and monitoring techniques might a production employee be engaged in the conception of at your facility?
   Example: Operators will assist in determining the interval for center-line monitoring during initial product runs.

6. What asset-related performance measuring and monitoring techniques might a production employee be responsible for executing at your location?
   Example: Operators will conduct and record Rockwell hardness sampling of every other production piece.

These six identified questions are examples of the ISO 55000 directed "control activities." Even if your organization chooses not to pursue an ISO 55000 certification, wouldn't operator engagement in control activities such as these make sense anyway?

It is time in our discussion on asset management that we understand a very basic, yet paradoxically complex element to any auditing process, ISO or otherwise. The idea is to present *objective evidence* that you are in compliance with your own processes. Over the years, I have conducted many audits across many subjects. The appearance of objective evidence is *always* the difference between pass and fail. And, within the 'pass' rankings, the more evidence of sustained adherence, the higher the mark.

I have been called into service to audit programs as far reaching as Process Safety Management (PSM) to Total Productive Maintenance (TPM) and everything in between. Whether an organization is working towards a certification for lean manufacturing or ISO 55000 Asset Management, one thing is necessary: say what you do, do it, and show that you have 'always' done it that way.

I don't care much for the practice of having two processes; one that we show auditors, and one that we actually follow. I can always tell the difference. The one that the auditor is shown is always the more pristine process documentation. That stands in contrast to the process that we actually follow, which is seldom documented at all. However, going forward, the process must be documented!

The training matrix that your organization will construct to ensure that the employees (operators) are engaged in the ISO mandated activities must show your organization to be one that has been adhering to their processes. If you're going for an Asset Management certification, you will need to compel auditors. If your organization isn't interested in the certification, you will still need the intricacies of a well-conceived matrix to serve you now to identify the gaps in your capabilities and into the future to close those very gaps.

The litmus test of any efficient and effective process is to say what you do, do it, and show that you have done it that way for a very long time. All three of these nuanced steps must be documented and unambiguous.

How would this look? Consider an example I provided earlier. I suggested that one asset-related performance measuring and monitoring technique that a production employee might be responsible for executing at my location was to conduct and record Rockwell hardness sampling of every other production piece. For objective evidence that this actual asset-related element is in practice at my location, I must have a process that clearly defines what this step means, show training certificates and the history of my operators performing this task, and demonstrate the sustained use of this process. I would have Rockwell hardness sampling records and test records, evidence of issues that might have been discovered, and some problem-solving casework documentation if necessary.

If you are imagining a very comprehensive training matrix, then you're thinking in the right direction. The matrix itself is just the visual manifestation of where the organization stands as it relates to gearing its operators up to perform at a higher level of execution. Figure 5-2 is an example of a training matrix that might be compiled to show evidence that our organization takes this challenge from ISO 55000 seriously and has put some joules of energy behind it.

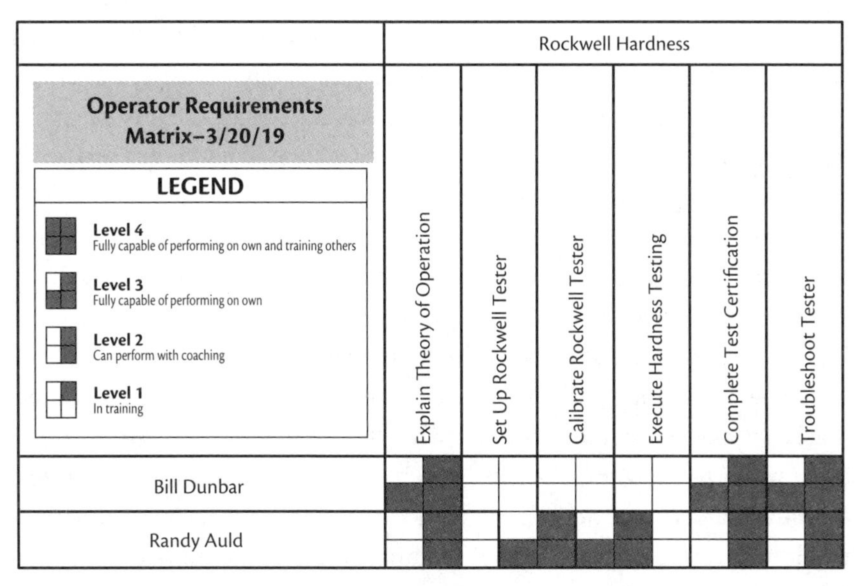

Figure 5-2 Operator requirements matrix
Adapted with permission by Marshall Institute, Inc.

For satisfactory objective evidence of this mandate, the matrix alone will not get the job done. The training matrix has an accompanying process guide behind it. The process guide describes the details for accomplishing each task. You saw this style of training matrix before in an earlier chapter.

All this effort will be little more than Kabuki theater if we stop here and don't challenge our operator corps to grow, get more involved, and be allowed to benefit, financially or otherwise, with the greater responsibilities they are assuming.

What's Next?

It seems that today, more so than any other time in industrial America, the need for a bridging point between production and maintenance is at its greatest. In a strange alignment of the 'stars,' we have arrived at this need and this point in time with a working class that is most suitable to cross that bridge. Our operators are smart.

I want to make a generational distinction. Our operators have always been smart and extremely capable. In fact, as operators, they were 'better' back then because operators seemed to be more in tune with their equipment. When I started in maintenance, an operator could tell me when something didn't *feel* right with the equipment. Today, that may not be the case, but then again, it isn't completely necessary. With the advent of the Industrial Internet of Things (IIoT), operators are likely to get text messages informing them that something is not right with their assets.

I have actually coined a theory about this phenomenon. I call it the *car wash theory*. In years past, people were more likely to wash their cars by hand, in their driveway on a nice spring evening while the kids were playing out in the cul-de-sac. By washing and waxing their cars by hand, the people—the *vehicle operators*—were becoming more intimate with their machines and more likely to sense when something was not right. Today, it is rare to see neighbors washing their cars by hand out front on a nice evening. Car owners are more likely to drive to a local car wash and let a machine clean their machines. Which of these two methods would net a more sensitive operator?

Our operators today might not be as sensitive to machine conditions. This is not an indictment against production and may not be applicable to your company. They (the operators) are, however, more knowledgeable and willing to participate in operator care programs. It is not unusual for me to hear an operator in a manufacturing plant say, "Give me a wrench and I'll fix it myself." Could we be better served as asset managers to tap into this interest?

About five years ago I was taking a tour through a manufacturing plant and a machine caught my eye. I asked my host, the maintenance manager, to stop for a minute as there was something unusual about this machine we were walking past and I wanted to get a closer look at it. What caught my eye was the fact that every fastener that I could see on the machine was sticking up about half an inch. Where I had expected to see the typical hex-bolt fastened down to the machine with only the head protruding, I found a nut welded to the top of each bolt head. This gave the machine the appearance that all the visible fasteners were sticking up. When I questioned the operator about this he told me that in order to change over this particular machine the action required eleven different combination wrenches. In order to reduce the time and the effort to change over the machine, the operator told me that he had welded the same size nut onto all the different bolt heads. Now he just needed one wrench to complete the task.

When I asked how the operator convinced maintenance to weld what appeared to be one hundred nuts onto the bolts he told me, "Maintenance didn't do it, I did it myself."

Please, somebody get that operator an apprenticeship!

This most certainly is not a new idea to many of you. The operator group is a built-in stable of potential maintenance apprentices. Of what benefit would it be to the asset management effort of the organization if the maintenance group was also comprised of some qualified operators? Imagine then if our maintenance apprenticeship program only drew from the top shelf operator corps.

This wouldn't be the current operator corps, per se, but rather one that had been engaged as a result of our adoption of Asset Management as defined by ISO 55000. We would have a production group that has squarely been engaged as stakeholders in asset policy and procedures. This group has, from onboarding, been heavily trained in the theory of operation and escalation of maintenance. This is a group that has been groomed to *wash their cars by hand* and have a handle on the technology of today's machines. We have a smart and capable operator group; what's next for those operators that seem to float to the top of the crop?

Can you name at least two current asset operators who would be ideal candidates for a maintenance apprenticeship program in your facility? Just as an exercise, write down the first names of those two individuals:

What made these individuals float to the front of your mind? Was it their attitudes or was it their capabilities? It might have been both.

Through the ideas of escalation of maintenance and a ready-made apprenticeship corps in the operating group, the lines between maintenance and production seem to blur or at least begin to gray. What's chiefly of interest is that we leave some clear lines of demarcation to keep our eye on the prize, which is the effective and efficient use of our assets in the accomplishment of the organizational objectives. ISO 55002 mandates that we have an asset management policy and that within that policy, responsibilities and the authority to execute aspects of asset care are clearly defined. There is a bridge that needs to be built between maintenance and production, but keep in mind, there is a bridge.

The Doctor Will See You Now

I am truly grateful to be in association with some remarkable professionals in the space of equipment maintenance and reliability. One of my colleagues made a point a few years ago during a discussion that maintenance is a *service* that should be *pulled* into production. In other words, maintenance should not be forced on production. My colleague asked, "When was the last time a car mechanic showed up at your house at 9:00 on a Saturday morning to change your oil, whether you wanted your oil changed or not?" In a world of lean manufacturing and lean maintenance, all work should be pulled through the organization.

I titled this section "The Doctor Will See You Now" because I want to demonstrate some parallel arguments of a routine practice, such as going to see the doctor and plant maintenance, very similar to the greater point my friend had been making.

ISO 55002 instructs the organization to determine the action plans and provide the resources to address risks associated with assets. Consider what we've been discussing in this chapter, and what role the operator and production at large have to play in asset management. In the future state that we have been building over these last few pages, the operator is an engaged and involved stakeholder in our asset management approach and as such, is required to actively participate in the care of the asset.

Much like making and keeping an appointment with your family doctor, it is imperative that there be a balance of responsibility between the two factions: the patient and the doctor, production and maintenance.

For either urgent or routine care, it is contingent upon the patient to make an appointment with the doctor in order to be received and to receive medical care. The patient is the driver of the process to get medical attention. For routine care, such as a physical or checkup, the patient has a better chance of being seen by the doctor at a time that is convenient for the patient if an appointment is made well in advance. If the care is more urgent, the patient may receive the care based on the priority of their individual situation. It is likely that the more urgent care issues will need to be addressed by a medical group associated with urgent or emergency care. For those circumstances, the care will be administered at the emergency center once more based on the priority.

In any case, the patient is required to go to the place where the medical attention is provided. The work can either be scheduled if the circumstances allow, or can be brought under the guise of urgent or emergency care if necessary. In any instance, one of the first measures collected is the vital signs of the patient. I've often wondered what the purpose of this process was, but I have been assured that the measurements are necessary to establish a baseline of the most basic health measures: temperature, pulse, blood pressure, lung function, etc.

In any and all scenarios, the doctor or nurse will ask either, "What are we seeing you for today?" or say, "Tell me where it hurts." Some exchange from patient to medical professional is necessary. You (the patient) have to tell the doctor what your concerns are. Medical care is rarely forced on an individual.

You might have assumed, and you would be correct in thinking, that I'm drawing a comparison between a well-known scenario of going to the doctor and the interchange between production and maintenance for asset care. More than just making a cheeky analogy out of this example, I want to stress that it is necessary to make it the policy of the organization that this is how the process is going to work.

Here is the comparison: in an operating facility, there will be occasion for assets to require attention either for routine needs or urgent needs. You can imagine that a routine need might be a preventive or corrective task, and an urgent requirement might be synonymous with a breakdown.

For a routine action, such as preventive or corrective work, much like a checkup with the doctor, production is better served by making an appointment to be seen by the doctor (maintenance) to ensure the need is addressed at a time that can be guaranteed and at a date that is most convenient for production. How much better is your personal health if you make an appointment with the doctor, in advance, and then don't go to the doctor to receive the care? That wouldn't be very helpful to your health, nor as it turns out, is it helpful for production to have time scheduled with maintenance and then not allow the care to be provided.

For urgent or emergency care the need for communication and frankness is even more pronounced. The doctor needs to hear exactly what is going on, and exactly what may have led to your particular affliction. The same can be said for emergency calls on equipment. The interest in both cases is to get the *patient* up and running as soon as possible.

An advantage to be gained from a formal process such as the adaptation of ISO 55000 is to hire, train, and bring to bear an operator corps that is intelligent and intuitive enough to help in their own diagnosis. Further, an advanced group of *employees* now having been regarded as critical stakeholders are better agents in managing the asset's health and contributing to a long life of service towards the organization's objectives.

This idea that maintenance is a service to be pulled into the production operation may be a new way of looking at this relationship for your organization. In order to advance together as partners in ownership of the asset, production and maintenance must share the asset for their particular use. Plant assets are either running or getting made ready to run. Those are the two typical states of plant equipment. In order for production to add value and to perform the tasks that they assume add value to the organization, the asset needs to be running. It should come as no surprise then that in order for maintenance to do their share and add value as they are requested, the equipment is usually not running.

How do we know when the asset should be running and when it should be down for maintenance and other servicing activities?

I'm Givin' Her All She's Got

The answer to the question of how long we should run a machine before we stop it for some maintenance or other servicing activity is simpler than you would think. I'm including in this text a short excerpt from some earlier work that I use to determine a suggested run time for production. (Ross, pp. 326–329) Work through this brief section and determine if it has some applicability to you at your location.

I promise you that this will be the most math intensive section of this entire workbook. In fact, you will need a calculator, pencil, and paper—or just write in the margins. What we are going to be exploring is a method that you can use very soon to make a compelling argument for a predictable maintenance schedule.

Ramesh Gulati's great work, *Maintenance and Reliability Best Practices*, includes a section on calculating the reliability of an asset and continues with calculating the reliability of a line of assets that are set up in series or parallel. We are not going to go into all of that, but rather, we are going to look at a singular reliability number and determine the maintenance schedule for that entity.

The very first time I did this process, I was at a corrugated box company out west. I had somehow gotten myself into a discussion, an argument really, between the operations manager and the maintenance manager.

The operations manager wanted to run a particular machine for twenty-four continuous hours. The maintenance manager was telling him that he couldn't do that. Now, I know for you and me, running a machine for twenty-four hours is nothing compared to almost everything in your plant. But this was the scenario taking place before me, and I was merely a fly on the wall.

The operations manager looked at me and said, "John, the Mean Time Between Failure for this machine is twenty-six hours. That means that I can run it for twenty-six hours without a breakdown, right?"

I told him that wasn't exactly what Mean Time Between Failure meant, and by the way, twenty-six hours was horrible!

I showed him the formula in Figure 5-3, which is slightly modified from Gulati (p. 163)

$$e^{-\lambda(t)}$$

Figure 5-3 Reliability calculation

As Gulati explains:

- e is the natural log. It is a constant value, which is approximately 2.71828.
- λ is the Greek letter lambda. In this formula it represents the inverse of the MTBF.
- (t) is the time that the asset is intended to run, based on the same units as the MTBF (hours, minutes, etc.).

What I sketched out for the operations manager was something similar to Table 5-1.

Table 5-1 The Client's Reliability Calculations

Intended Run Hours	**λ**	**Resulting Reliability**
24	0.0385	.39 or 39%
12	0.0385	.63 or 63%
8	0.0385	.73 or 73%

I told the production leader that with a MTBF of twenty-six hours, he had about a 39% chance of getting through a twenty-four hour straight production run without a hiccup. In fact I said, if I were going into a production run, I would want a reliability of greater than 85%.

Test your proficiency with the formula in Figure 5-3. I'll provide the scenario:

Your A-1 packaging machine is a stand-alone asset and is used sparingly during off-peak production and more often during peak production. You and your team are confident in your calculation of the MTBF of this equipment, and the number you have arrived at is 560 hours.

The MTBF is 560 hours. What is the inverse, or lambda?

The production team wants to run this packaging machine for thirty straight days, around the clock. This is the peak period.

What is the resulting reliability of that scenario?

I realize that I am using reliability, probability, and chance interchangeably, but the synonym works for this scheduling discussion.

You have calculated the lambda, given the MTBF figure, and you have calculated the reliability figure using the formula in Figure 5-3. Now calculate what continuous run time would be permissible to have a resulting reliability (probability) of getting through that run time of 85%. What is your answer?

My answers:

Lambda = 0.0018 (the inverse of MTBF)

The probability of getting through 30 days = 27%

For a probability of 85% = run for 90 hours at a time, or 3.75 days.

Now hold on, this is the interesting part and how this all ties into the schedule. With the numbers as just calculated, using mine or whatever your values are, Figure 5-4 would be the proposed schedule approach to this particular machine:

90 hrs. running	1 hr. maint.	90 hrs. running	1.5 hrs. maint.	90 hrs. running	1 hr. maint.	90 hrs. running	2 hrs. maint.	90 hrs. running

Figure 5-4 Proposed run schedule for the A-1 packing machine

Based on the historical run record of the A-1 packing machine, our numbers tell us that we can run for ninety continuous hours with a *probability* of success (no unplanned downtime) of 85%, or perhaps greater. This is a very loose, yet I believe a practical way to use historical data to determine how often we should perform maintenance on an asset.

The maintenance times shown in Figure 5-4 are just examples of downtime periods. The numbers you would use would be the real numbers. The question now becomes, "what do we do during those downtime sessions?" I suggest we perform the required PMs and corrective planned maintenance. We really want to do what a colleague of mine said many years ago: "just do something awesome."

As unique as it might be that production begins a process by which they pull maintenance into the operation, it seems practical that production has some idea of when to call for maintenance. If we couple that thought with a further desire to be proactive, or to fix something before it begins to break, it makes sense to run with a probability of success of >85%, stop and perform maintenance, run again, stop, and so on. Do something awesome.

To round out this discussion on production's role in an asset management environment, consider an earlier analogy of a doctor and a patient. The medical care and service rendered will never be better than when the patient and the doctor are working together, communicating and each aiding the other. The health of our assets will never be better than when production and maintenance are working together in that same spirit.

CHAPTER SUMMARY

We can't afford for asset management to be synonymous with maintenance management. Not only would that be disastrous but would set back the clock on autonomous maintenance and all the advancements our operator groups have made over the decades. Our operators are just as capable and knowledgeable today as they were in the past. Albeit, it is a different time, but we have within our production group a ready army of stakeholders mandated by

ISO 55000 to be part of the planning and execution of organizational objectives relating to the company's assets.

Opening planning and policy development opportunities to unknowing and unwitting employees would be a disservice to the company and more importantly, to the employees themselves. Rather, let's consider what is most desirable to achieve in terms of actions and resulting benefits of a trained, engaged, rewarded, and enterprising operations partner.

Adding to Our Business Case

It seems that during the onboarding process an employee is likely to get more instruction on how to fill out a 'vacation request form' than they do on how to fill out a 'work request form.' That needs to change. Odds are that an associate will need to request work from maintenance more often than they will vacation days from their boss.

This must be a cautionary tale however, as the answer is never as simple as showing operators how to fill out a work notification or work request. An operator must first have a sense of the theory of operation of their equipment before they can then determine that something is wrong with it.

At our facility, operators are provided theory of operation training by:

Our asset management policy includes a notice that operators and other production affiliated personnel will have a robust onboarding process that includes an extended theory of operation. The onboarding process for production associates should include an unambiguous exchange of the new employee's responsibilities as they relate to the organizational objectives, their responsibilities toward the care and operation of their assigned asset(s), and a very comprehensive plan to measure and assess their knowledge and development of the training protocol to close the gap of their current understanding and the company's expectations.

To exercise that knowledge, operators and production personnel will become an integral part of the overall maintenance and reliability effort through a formal and very well-defined escalation of maintenance protocol.

Currently, the steps at our facility to communicate equipment issues flow like this:

Our efforts to hone the skill of certain operators to a higher level of capability and asset care will be enhanced with the addition of an operator-tech position allowing for more focused delivery of the skilled trades where needed. Inputting the role of an operator-tech, the maintenance escalation process would now progress through several in-line skilled and trained operational human assets before diverting maintenance human resources from their planned and scheduled work:

- From the Operator to his or her Supervisor
- From the Supervisor to the Operator-Tech
- From the Operator-Tech to the Maintenance Technician

Production's role in an asset management world takes on a new set of requirements and stretches the employees into roles that they have been traditionally not engaged in; namely the conception of asset-related policies and procedures. Additionally, the operator group must actively participate in the development of measures and performance indicators to determine if the activities surrounding an asset are in line with the organizational objectives. This is a significant role addition and, as such, an equally significant effort will be taken to ensure that our operator base is educated and equipped with the knowledge and skills to be major contributors to the discussion.

For their role in asset-related policies and procedures, production personnel will be engaged in formal training, which will be documented on a skills matrix similar to the one shown next:

Operator Requirements Matrix–3/20/19

LEGEND

- **Level 4** Fully capable of performing on own and training others
- **Level 3** Fully capable of performing on own
- **Level 2** Can perform with coaching
- **Level 1** In training

	Rockwell Hardness					
	Explain Theory of Operation	Set Up Rockwell Tester	Calibrate Rockwell Tester	Execute Hardness Testing	Complete Test Certification	Troubleshoot Tester
Bill Dunbar						
Randy Auld						

The increased role operations are to play in the asset's health doesn't stop at this intermediate position. Highly trained and capable operators can now be considered ideal candidates to fill the greater need in the skilled trades through a formally devised and executed apprenticeship program. Selection of successful candidates is paramount to the health of the organization and the continued delivery of asset performance. Imagine a core of exceptionally capable trades people who have 'come up' through the operator ranks to bring a well-rounded wealth of operational and technical knowledge to future asset discussions.

Although the lines of separation between maintenance and production begin to blur and gray around a high performing asset management company, each skill has its place and its time. With that in mind, production will assume the necessary role of pulling maintenance into production to be in line with the synergies of lean manufacturing. It is necessary for production to establish the level of care they wish for the company assets to deliver to ensure the needed performance matches the organization's objectives. Maintenance cannot be forced on production and each side must determine the appropriate time for scheduled maintenance.

The time estimate for scheduled maintenance in our facility will be derived from asset performance history and utilize the reliability formula shown here:

$$e^{-\lambda(t)}$$

- e is the natural log. It is a constant value, which is approximately 2.71828.
- λ is the Greek letter lambda. In this formula it represents the inverse of the MTBF.
- (t) is the time that the asset is intended to run, based on the same units as the MTBF (hours, minutes, etc.).

Maintenance will provide the best technical advice and it will always center on a system reliability value of >85% using the formula just shown. As an example of this formula in use, consider this scenario:

An A-1 packaging machine is a stand-alone asset and is used sparingly during off-peak production and more often during peak production. You and your team are confident in your calculation of the MTBF of this equipment, and the number you have arrived at is 560 hours.

The MTBF is 560 hours, which means the inverse, or lambda is:

The production team wants to run this packaging machine for thirty straight days, around the clock. This is the peak period.

The resulting reliability of that scenario is:

The continuous run time that would be permissible to have a resulting reliability (probability) of getting through that run time of 85% is:

Using this scenario, the resulting run schedule as advised by maintenance would look like this schedule shown:

90 hrs. running	1 hr. maint.	90 hrs. running	1.5 hrs. maint.	90 hrs. running	1 hr. maint.	90 hrs. running	2 hrs. maint.	90 hrs. running

Turn to the last section of the book and record this same information in the comprehensive strategy section.

SIX

Asset Management at Your Place and at Your Pace

"Watch out for organizational inertia. It is natural for organizations to resist change.... Remember, however, that you should never be overly constrained by the existing organization."
—Bert Nanus, *Visionary Leadership* (p. 169)

I resisted, quite successfully I must add, putting the word "maintenance" in the title of this last chapter. I should stress again that ISO 55000 is not a maintenance program, nor should you attempt to make it one. The standard clearly states that the organization is responsible for identifying the stakeholders who are important to asset management and that their needs and expectations will be considered. Those stakeholders include employees and other such internal groups, including but not limited to: engineering, accounting, maintenance, and operations. This is not a maintenance program. I chose the Bert Nanus quote at the beginning of this chapter explicitly to stress the point that it's not business as usual for companies working to adopt ISO 55000 and all the tenets there within.

In this final chapter we are going to walk through the standard and highlight some of the more salient points and work together to provide the type of objective evidence you will need to compel others as part of a certification in ISO 55000. Your organization may not be interested in a formal ISO certification, yet you may simply want to benefit from the structure of this international approach. This chapter will be helpful for those companies as well. There is one thing I'd like for you to keep in the back of your minds as you work through Chapter 6. Frequently ask yourselves the rhetorical question, "Does this make sense?" I believe that you will find, even if your company is not interested in formal certification, that an official policy and program covering asset management is the right action to take for the future of your organization. The challenge in any case is to make asset management work at your place and at your pace.

We are going to focus on several key elements that are necessary to successfully manage assets. I will be using and interpreting ISO 55000, 55001, and 55002 for reference and bringing in other text as it supports the argument. Again, throughout this chapter, continue to ask yourselves if this all makes sense. I believe that you will agree with me that this approach makes perfect sense. Our discussion in this chapter will center on:

- Organizational plan
- Organizational objectives
- Asset management policy
- Strategic Asset Management Plan (SAMP)
- Asset management objectives
- Asset management plans
  - Operating planning and control
  - Supporting activities
  - Control activities

More detail is required on these bullet points, and partway through the in-depth review of these topics, I will interject a scenario meant to drive the remainder of our discussion towards *physical* asset management, a focal point I stressed at the very beginning of this book.

Organizational Plan

The organizational plan is the corporate plan. That sounds simple enough, but consider that the plan speaks to how all this is going to actually work. Those drafting the organizational plan must be conscious of the context in which the organization performs. This might be challenging if you can imagine a large organization with operating facilities around the country or the world. Each plant might operate in a different context. By context I'm referring to operating environment, local and federal laws, unions, and community interest among other aspects defining the circumstance in which each facility operates.

The organizational plan is the critical, necessary, and vital first step to an asset management process. Did I overstress the importance of the plan? An organizational plan includes great detail about how the company intends to be successful. These 'details' are the objectives, and organizational objectives are relevant to what we are discussing. These objectives come from the strategic planning accomplished at the highest levels of the corporation. Without this roadmap, all our efforts to work together to produce goods and services, at a sustainable cost, and in an attempt to beat the competition, will be reduced to just a bunch of good people fighting the good fight.

The objectives within the organizational plan set the tone and the character of the organization. The organizational plan, or corporate plan, should be a greatly communicated idea that resonates with all the stakeholders of an enterprise. Recall that stakeholders include company employees, contractors, government officials, and customers. The plan should be communicated in such a way that all parties are knowledgeable as to what the organization is all about and the impact it hopes to have in the space in which it operates. It may need mentioning that the company is not obligated to share trade secrets or sensitive financial information. However, there should be objective evidence to demonstrate the connection between the organizational plan and the effort to engage those stakeholders.

Keep in mind as we start to detail out how to actually develop a culture and sense of asset management in your organization that the leadership is responsible for determining the scope of the asset management system. The Strategic Asset Management Plan (SAMP) should capture the boundaries and applicability of the asset management system. There is a two-way link between the organizational plan and the SAMP. Can you imagine the result of an organizational plan and detailed objectives that in turn ran counterintuitive to the asset management plan? I believe you not only can imagine it, but that you might actually be living it right now!

Are you aware of your company's organizational plan, the big picture? If you are able to articulate any aspect or feature of that plan, try to document those segments in the space provided:

Organizational Objectives

There is a reason that organizational objectives are some of the very first elements of an effective management system. Don't assume you have an asset management system if you don't have a well-established and communicated series of organizational objectives. Just don't do it.

I have mentioned many times throughout this text that the stakeholders have a mandated engagement requirement in the process of asset management and specifically in the establishment of what the organization is trying to accomplish. This is for good reason. Here are a few traditional organizational objectives that I'm introducing as examples of what you might find in your own organization.

- Profitability
- Productivity (of people and resources)
- Customer service
- Core values
- Growth (sustainable)
- Cash flow
- Change management
- Beating the competition

These objectives as well as others help to define the trajectory of the organization and effectively point the direction the company must follow to ensure that all its assets (physical, human, financial, etc.) are working together (in harmony) to reach the goals of the organization.

Of course, I don't know your specific circumstances, but if your organization adopted the practice of engaging its employees in the development of these objectives, would you personally have something to add to the conversation? One common shortcoming to leadership engaging the workforce in decision making is that if it's never been done before, the employees may be initially shocked into silence. Is your organization guilty of this?

What follows is a little give-and-take exercise that I'd like you to complete to determine if you, personally, are engaged in the organizational objectives as an employee. I am using the organizational objectives I just listed as examples. I certainly would understand if these objectives are not the objectives of your actual company. This is just an exercise.

Making a Profit

As a general example to test how much communication there is within your own organization, use the space provided to explain how your company intends to make and sustain a profitable business.

Companies maintain a profit by selling goods or services for more money than they cost the company to produce. That sounds simple enough, but it's often hard to pull off considering there is so much that goes into producing a product or service. There are often numerous variables and even access to raw materials is questionable given certain trade restrictions and governmental obstruction.

The focus for staying profitable should be on controlling costs while protecting the profit margin on products and services that are sold. Imagine running a business where your costs for producing were not controlled and as a result, you were constantly increasing the price of goods to your customer. How long do you imagine you would be in business?

For the sake of objective evidence that the element of *profitability* as part of your organizational objectives was met, please document in the space provided the last business financial meeting you attended and also document how you were personally asked to be engaged in the conversation.

Companies maintain a profit by providing goods or services for more money than they spend making the goods or service. This *delta* in costing is the profit, or profit margin, alluded to in Figure 1-7.

Keeping It Productive

This is the exact organizational objective element where the execution of maintenance is front and center. If you will consider Chapter 5's insistence that production become a greater player in the reliability and upkeep of assets, then you will agree with me that 'maintenance' doesn't necessarily mean the maintenance department.

In order to make production, maintenance, and all other stakeholders into productive members of the organization, it would stand to reason that there is a training program commensurate with the effort needed to make them 'productive.' Does your organization have a robustly documented and executed training program similar to the programs outlined in the preceding chapters?

It is not enough to just provide the training and the training platforms to keep associates moving to gain greater knowledge and abilities. Under this particular element, the organization must also resource the employees with the tools, time, and direction to execute the jobs they are assigned. In the space provided, briefly explain how your organization resources (or equips) its employees to perform their assigned duties.

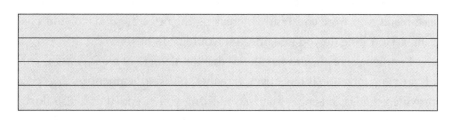

As a final thought, and one that requires your input, an organization must be able to measure the productivity of its endeavors. If the intent is to improve on productivity, doesn't it stand to reason that there is a means to measure productivity in the first place? At its most basic, productivity is a measure of output over input. How does your organization measure productivity for the following departments?

Production
Maintenance
Engineering
Accounting
Human Resources
Safety

The Customer Is Always Right

Do you think that's true? Is the customer always right? I worked in an ice cream and dairy store for a few years when I was a teenager and I can tell you that sometimes the customer didn't even know what they wanted. The actual focus for this element is customer service and customer satisfaction. There is a popular business management axiom that it is cheaper to retain a current customer than to find a new one. If you subscribe to this belief, and you should, would you also agree that a happy or satisfied customer will be one that will promote your business to other potential clients? Word of mouth is a powerful advertising tool.

I had a client once that was insistent on using a particular phrase in their vision statement. The phrase that had to be in their reliability vision was "to elate our customers," or some version of the word 'elate.' When I inquired as to the origin of this unique phrase I was told that the owner of the company was passionate about elating their customers. "Let our competitors *satisfy* the customer," the owner was fond of saying, "We're going to elate ours!"

Excellent and superior customer service should be the hallmark of any organization. Can you recall an episode where, as the customer of some product or service, you were not elated or even satisfied? How did that make you feel? If you were forced to use that service because of limited competition, I would bet you probably felt trapped and very unhappy about your lack of choices.

What objective evidence is there that your organization is continually working to provide excellent customer service? Record your thoughts next.

At Its Core

The mission statement that guides your company should include a description of what your organization holds as its core values. What are the core values for your company as they have been communicated to you? Also, what objective evidence is available to show that these core values are known to all the associates and that the company lives by them?

Core values:
Objective evidence:

A company's core values often address the beliefs the company has towards customer relations, its responsibilities towards the community and environment, and how the organization works to create a positive culture among its associates and other stakeholders.

If You Ain't Growin' You're Dying

I have attributed this commonly used phrase to the legendary Notre Dame football coach, Lou Holtz. Unless I hear otherwise, I'm going to continue to think that. The issue for discussion as it relates to organizational objectives is the one of sustainable growth. As you might imagine, the future growth of the company must involve a detailed discussion on how the company assets, both human and capital, can support a growth strategy.

The *sustainability* attribute of growth has to remind leadership that the abilities and capabilities of the company need to stay ahead of the actual growth commitment. This is rarely ever the case, and I don't suspect much will change because of this book. But it is my hope that companies will begin to think strategically when considering their assets and the execution of their business plans.

What process can you share as objective evidence that your organization practices the strategic methodology of sustainable growth? You may want to restrict your response to just those thoughts around capital assets and perhaps the people associated with the equipment. Please record your thoughts here.

It's All About the Benjamins

No one should doubt that the organization is in business to make money (refer to Figure 1-7). Maintaining a healthy cash flow is paramount to staying in business. A healthy financial position for companies allows those organizations to finance the capital equipment needed for growth and to better facilitate the flexible strategy required to outpace the competition. It doesn't hurt the image of the company to pay the bills and make payroll either.

In what manner does the organization invite the stakeholders to share in discussions on cash flow positioning?

The More Things Change, the More They Stay the Same

Change management is an often overlooked, and frequently poorly executed objective of most organizations. Can we be honest? What seems to make a lot of sense in the company boardroom is rarely accepted at face value in the plant breakrooms. I wonder how much the resistance to change is a result of employees not being in the *know* as it relates to where the company is heading.

In my first book I introduced a diagram that illustrated how culture changes in an organization. There are highly defined and required stages that must be appreciated in order for a change in direction to correlate to a change in attitude or culture. Review Figure 6-1, reprinted from *The Reliability Excellence Workbook: From Ideas to Action*, to demonstrate the flow of change through an organization.

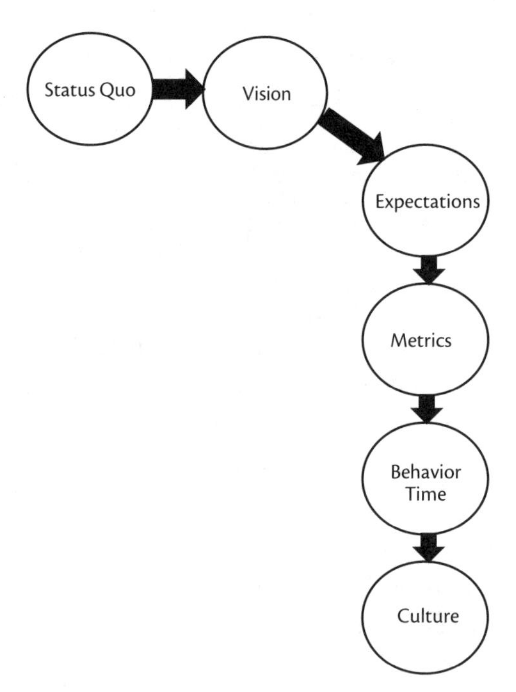

Figure 6-1 Change model

Some of the segments of the change model in Figure 6-1 have already been addressed in this section. But, just for review, let's walk through them again and look at them through the prism of change management.

STATUS QUO

No organization and no one in an organization is likely to move with earnest if they are comfortable with the way things are currently. In order to entice a group or an individual to move off the *dime* they are sitting on, there must be some discomfort with the as-is situation.

VISION

Of course, you can't just turn up the heat with where people are now and leave them there. As a leader you have to show them a better way. That is the vision. Leaders should never pass up a chance to get before their people and describe the *better place* that the company is heading.

The vision is a manifestation of the potential for the organization and has to be communicated clearly to show that everyone has a significant role in the future state. A leader shouldn't expect associates to be excited about a future state if their roles in the new place are unclear.

EXPECTATIONS

The leader uses the vision and the future state to set new expectations. It is here that the leader challenges the organization to reach new heights and accomplish things they never thought possible. Identifying, communicating, and accepting new expectations is vital to an organization keeping its edge, staying dynamic and being ready to meet the challenges of capturing opportunities, and growing the organization.

METRICS

Metrics drive behavior. Don't believe me? See if this scenario fits you. If you're driving down the road and a police vehicle is coming towards you in the opposite direction, do you look down at your speedometer? After the police officer passes in the other direction, do you look in your rearview mirror? Why do we do this? Because the metric (speed limit) is controlling our behavior. Also, we might be feeling a little guilty.

Where do metrics come from? Metrics are born from the expectations and are used to set milestones to determine if we are moving in the right direction. We trend the pattern of our movement and right the course when necessary.

BEHAVIOR OVER TIME

We don't change behavior, behavior is changed over time. Consider that one definition of behavior is "a general acceptance of new norms." To be sustainable, this acceptance requires a period of time to pass and must feel organic to the organization. Sometimes the time required for this acceptance can be quite extensive.

Charles Darwin spoke of change or variation that is "profitable to itself" as being the only sustainable change. The change 'happening' to the associates has to be change that they feel is beneficial to them. Certainly the change Darwin spoke of was significant and required lifetimes to achieve. We don't have lifetimes, but the need to progress slowly and purposefully towards our future is ever present.

CULTURE

The culture is the new place that an organization hopes to arrive at in due time. The goal of change management ought to be that the critical mass (the majority) of the associates arrive at this new culture around the same time. It is from this new culture position that *continuous improvement* efforts can be attempted.

Organizations have tried to adopt a principle of continuous improvement but never wait until the previous improvement attempted proves to be an improvement at all. Continuous improvement is predicated on the fact that the previous action was an improvement. Figure 6-2 shows the relationship between culture and continuous improvement.

What is truly stunning about the culture element and the change models in Figures 6-1 and 6-2 is how many leaders attempt to wrestle the organization towards a new culture, skipping all the steps between 'vision' and 'behavior over time,' or think that continuous change is continuous improvement. This does not bode well with the workforce, and the leader is likely to get compliance from the employees rather than commitment.

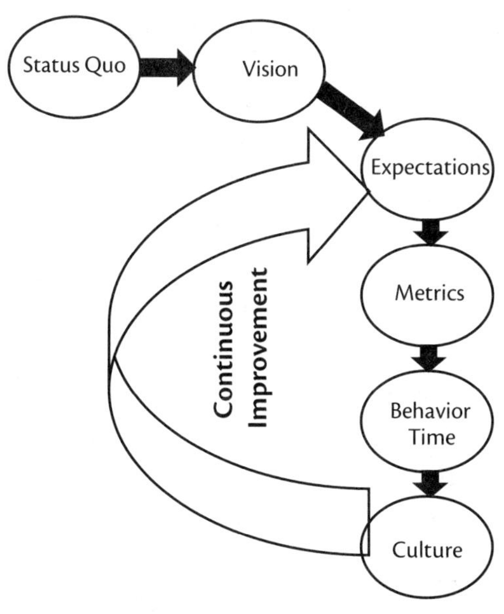

Figure 6-2 The relationship between 'culture' and 'continuous improvement'

The Beat Down

Staying ahead of the competition is a healthy and necessary organizational objective. From the preceding points you can conclude that a lot could go wrong. An organization would have greater success of outpacing the competition if everyone would work in concert to make the most out of their resources and persevere in their financial doings.

Before we can defeat our enemies we have to name them. In your industry, who do you see as your chief competitors? List them in the space provided.

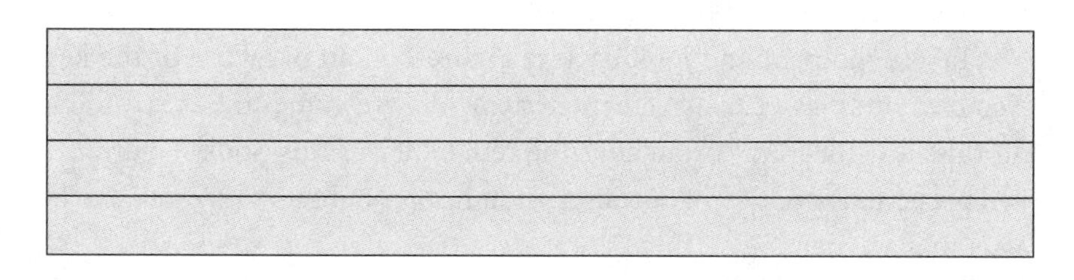

A large part of getting 'one over on your competition' is to take a hard look at how the company currently does business. It is vital that companies conduct a healthy examination of their organization and understand where their products and services sit in the marketplace among their peers. This process of critical self-examination can help to articulate a well-defined status quo (see Figure 6-1).

The organizational objectives should be far-reaching and lofty, yet speak to the common man. ISO 55000 intends for your organization to take the conversation of these core objectives down to the floor level, to the common man, if you will allow the analogy. Your organization might well be on their way to operating in this highly self-actualized state, but many companies still hold much of the information we've discussed close to the vest.

Can you list your company's organizational objectives? If so, list them here:

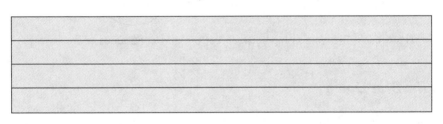

If you have filled in the 'blanks' so far in this chapter, you have at least made an attempt to list your company's organizational plan and the organizational objectives derived from that plan. Consider a scenario where a maintenance or reliability leader (or for that matter a production or plant leader) has no precise knowledge of the organization's purpose and intent. How would you know if you are doing the right things and leading your people in the right direction? I suspect, if you're being honest, many of you find yourselves right there, right now!

Take a moment and look back at Figure 2-6, an overview of the key elements of an asset management system. We are going to be expanding on this flow diagram throughout the remainder of this book. Figure 6-3 is representative of where our discussion has brought us so far, relative to asset management.

After creating and communicating the organizational objectives, work can begin on the creation of the asset management policy and the SAMP.

Figure 6-3 Relationship of organizational plan to organizational objectives

Asset Management Policy

Similar to any policy, an asset management policy is a statement of intent. This intent should address the guiding principles that are important to the organization and be supportive of the organizational plan and the organizational objectives.

Remember as you proceed through the text on asset management policies that you need to have a solid documentation trail and always be mindful of what objective evidence would support your adherence to each step. For an asset management policy, you might consider evidence that clearly indicates that:

- The policy was created by stakeholders
- The policy is followed
- The policy is reviewed from time to time to ensure it still supports the organizational objectives

The policy instills the organization's commitment to decision making and activities regarding asset management. For example, an organizational objective to make people and resources more productive could result in an asset management policy to report on asset and asset management performance.

Other policy principles might include, but are certainly not limited to:

- Following all applicable laws and regulations
- Integrating plant processes into corporate processes (e.g., budgeting, safety, business planning)
- Continuous improvement as it relates to assets and asset management

The asset management policy is a short statement and the details should be expanded on later in other documents. For now, these are just the guiding principles governing the organization in terms of its assets and the asset management system. Knowing and understanding the guiding principles can contribute to a feeling of confidence when associates make asset-related decisions and asset-centered contributions throughout the facility.

For objective evidence, consider chartering a small team to develop and communicate the asset management policy. Document the minutes and the policy, and create a suspense file to semi-annually review the policy for alignment with the overall organizational objectives.

For example, one of the organizational objectives written in the previous section was to make people and resources more productive. I also mentioned that if an organization wishes to make productivity an objective, they needed a method to measure productivity. Keep that in mind as you read through the next few paragraphs, which I have provided to show you what an asset management policy might look like.

An Asset Management Policy (Example)

PURPOSE

This policy outlines the principles used to guide decision making on asset management at the ABC Company to ensure the ABC Company meets its mission to provide high-quality products in a sustainable and safe environment.

SCOPE

This asset management policy applies to all production-related assets owned by ABC Company and is not applicable to research and development assets nor the headquarters facility assets. For the production-related assets, this policy includes the design, construction, operation, maintenance, and disposal of that equipment. This policy applies to all employees. We will work collaboratively to engage all stakeholders in the application of the principles outlined in this policy.

INTENT

ABC Company creates uniquely valuable products for our customers that require process ownership and engagement in the operations and maintenance of physical assets. The intent of this policy is to inform and ensure commitment of all ABC Company employees to the organizational goals as they relate to asset management to deliver on asset integrity and availability. And to do this in a manner that reduces risks and provides an exceptional level of service to customers in a safe and sustainable environment.

POLICY STATEMENT

In managing ABC Company's assets, we commit ourselves to:

- Being mindful of and following all applicable laws and regulations in the states and countries in which we operate and sell our products
- Integrating plant processes into corporate processes (e.g., budgeting, safety, business planning) to make product manufacturing seamless and trouble free at the plant level
- Continually improving as it relates to assets and asset management regarding equipment design, build, and install, and to have asset documentation and spare parts in place before asset start-up
- Making people and resources more productive
- Elating our customers

APPLICATION OF POLICY

ABC Company will develop and execute plans to guarantee that physical plant assets are capable of performing to the level commensurate with the pledge to our customers. This includes:

- Developing a long-term capital asset plan and adhering to the wisdom of the capital asset committee's strategy for asset replacement and modification

- Continually tracking market measures and making smart adjustments and corresponding updates to capital equipment strategies to match the trends
- Creating productivity measures and standards, applying those measures, finding opportunities for improvement and growing physical asset capabilities
- Analyzing the desires of our customers and engaging our associates to create energy around assets to exceed their requests

ROLES AND RESPONSIBILITIES

The roles and responsibilities for executing this policy include the following:

- The capital asset committee is tasked with developing and maintaining an extended capital asset plan to articulate the need to modify, buy new, or otherwise enhance assets to meet the changing production demand.
- The marketing department is responsible for keeping customer purchasing trends active and anticipating the need for changing employment of company production assets.
- Department leaders are required to determine appropriate productivity metrics to track, trend, and improve capital asset performance and availability.

POLICY ADMINISTRATION

Effective Date: 1 January 2019
Policy owner: John L. Manager
Application: All policies relating to ABC Company assets and personnel
Last Review Date: 1 January 2018
Version: Revision 002
Published Externally: Yes
Approved By: CEO Johnson

Objective Evidence

The objective evidence that your organization has an asset management policy is that you actually have one. No mystery there. As mentioned earlier, this policy is ripe for constant review as corporate objectives and plans are often changed or modified. Organizations may change strides in the middle of a race and not fully communicate that to the operating facilities. Better said, this information may make its way to the plants, but it would not be uncommon for an asset management policy to be at odds with the new corporate direction. The result would be an asset management policy that is 180° out of sync with the organizational objectives.

Does your organization have an asset management policy? If so, state your policy here:

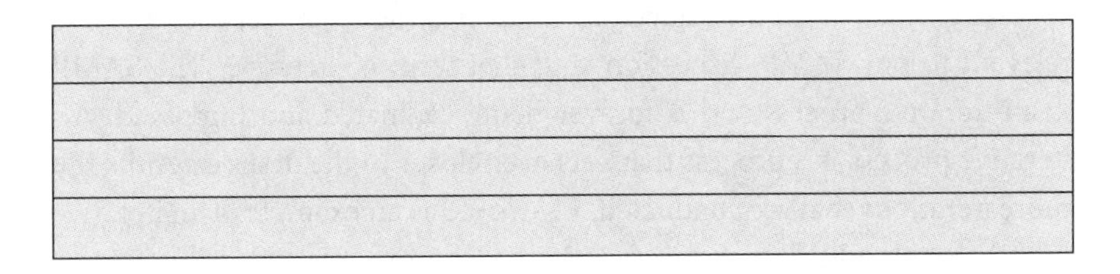

Figure 6-4 shows where we have progressed in our discussion on asset management so far.

Figure 6-4 Asset management policy from the organizational objectives

That asset management policy and the organizational objectives merge at the Strategic Asset Management Plan (SAMP).

SAMP

The SAMP is the transformational document that converts organizational objectives to asset management objectives. A well-conceived Strategic Asset Management Plan will also lay out the approach to be taken to comply with the overall asset management objectives. Before we get too far into the SAMP, I want to stress again the importance of creating and communicating the organizational plan. A quick review of Figure 6-4 will confirm that you might have trouble with a SAMP if you don't first have a solid organizational plan and well-communicated organizational objectives.

There was a term introduced in Chapter 2 that bears repeating and is relevant to our SAMP discussion, and that term is 'iterative.' The SAMP is an iterative process and is forever being evaluated and improved. An iterative process is a process that gets ever closer to the desired results the more iterations that are conducted. Figure 6-5 is an example of an iterative process we should all be familiar with: the Deming Wheel.

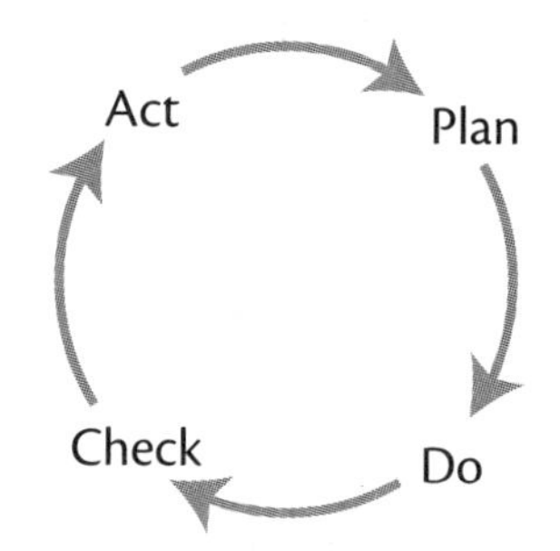

Figure 6-5 The Deming Wheel

As an iterative process similar to that shown in Figure 6-5, keep in mind that your organization needs to show that there is constant reviewing and aligning of the SAMP and asset management objectives with the organizational objectives. These reviews become the objective evidence that the SAMP is a living product and is 'freshly' in tune with the direction of the company.

You might have picked up on a new term in that last paragraph. If not, here it is again: asset management objectives. What is an asset management objective?

Asset Management Objective

Asset management objectives are the desired goals (targets) for the assets and the asset management system. Some examples might include:

- Achieve 90% production asset availability.
- Achieve 98% PM compliance.
- Increase asset and component reliability (increase Mean Time Between Failure) by 10%.
- Increase asset and component maintainability (decrease Mean Time To Repair) by 10%.
- Achieve 3% maintenance cost reduction (from the previous year's budget).

As these are only goals or targets, there is no indication as to 'how' the desired results are going to be achieved. The asset management objectives that your company seeks should be supported by a greater initiative, or the means by which the goal is met. This is fundamentally accomplished as part of a Goals–Means–Consequence effort. For example:

- **Goal.** Increase PM compliance.
- **Means.** Conduct Preventive Maintenance Optimization workshops, create a dedicated PM crew.
- **Consequence.** PM compliance increases to 98%.

These initiatives, or 'means,' can be part of a larger effort by the organization, or something more local to drive the improvements sought. The 'aha' moment you should be experiencing right now is how in Chapter 6, so far as an example, we've connected the:

- *Organizational Plan* of engaging the stakeholders to make the company 'successful' to the

- *Organizational Objectives* of profitability, customer satisfaction, and productivity, to the
- *Asset Management Policy* of creating long-term asset management plans, elating the customer, and making people and resources more productive, to the
- *Asset Management Objectives* of increasing PM compliance, reducing costs, and increasing asset reliability and maintainability

Honestly, for you and me to proceed any further in this book, I need for you to be able to see how all of these process steps are connected to and supporting each other. That is the *strategic* part of asset management. Look at those bullet points just listed and read them backwards. Can you see how accomplishing all that is in the first (last) bullet leads up to what can be done in the next bullet, and so on? This has got to be constantly and consistently communicated throughout the organization.

In the space provided, list any asset management objectives that you can show documentation for:

Documentation

The SAMP is a documentation tool. One area clearly served by this documentation means is to link the strategic initiative and the asset management objective. The recent Goal–Means–Consequence linkage offered as a recent example demonstrates this benefit. In order to achieve the goal of increasing PM compliance, the company has to put forth the resources (read that as time and money) to make this happen. Those resources manifest themselves as PMO workshops and as creating a dedicated PM crew; these are the 'means' to accomplishing the goals. The SAMP becomes a method for documenting these links.

The quality of the SAMP tool is measured by how well the objectives and initiatives are communicated. Recall that the associates and indeed all stakeholders are part of the council in developing the asset management system. Here in the SAMP, the employees learn of how well those initial thoughts and their communicated ideas were heard and heeded. The communication 'out' from the SAMP is predicated on having the right information, aimed at the right people, and delivered at the right time.

Requirements

Understandably, and actually to our favor, the ISO 55000 standards leave a lot for us to interpret as it pertains to what a Strategic Asset Management Plan actually is. Aside from the organizational objective-to-asset management objective connection we've just made together, there is very little in the way of direction that is not vague in nature. In fact, the standards appear to employ a level of circular logic to explain what is needed in a Strategic Asset Management Plan.

I said that this is to our advantage because like most ISO applications, we simply need to document our interpretation of the standards and execute what we believe to be compliant. Have the follow-through and follow-up objective evidence to prove that you do what you say that you do.

There are a few actual references to SAMP in the standards. The following SAMP directives are extracted from ISO 55002:

- Section 4.1.1.1 states that the approach to implementing the principles set forth in the asset management policy should be documented in the SAMP and goes on to inform us that the SAMP documents the connection between the organizational objectives and the asset management objectives and that the links between the SAMP and organizational plan should be a two-way and iterative process.
- Section 4.1.1.2 mandates that when developing the SAMP, that the organization should:
    - Give thought to the expectations and noted requirements of the stakeholders.
    - Be mindful of the activities beyond the organization's planning time frame, and those activities should be reviewed regularly.
    - Clearly document the asset-related decision-making criteria.

- Section 4.2 suggests that the stakeholder's statement of needs might be documented within the SAMP and should make reference to required items and issues (recall that government agencies are stakeholders).
- Section 4.3 indicates that the SAMP should establish the boundaries and applicability for the asset management system; this is the scope.
- Section 4.4 advises us that the SAMP is created after the asset management policy.
- Section 5.2 further advises us that the asset management policy could be included in the SAMP.
- Section 6.2.1.1 informs us that the asset management objectives are derived from the SAMP.
- Section 6.2.2.1 mandates that we look for gaps in the ability of the SAMP to support the achievement of the asset management objectives.
- Section 7.2.2 indicates that the SAMP is to be considered when performing a gap analysis on the alignment of the asset management system.
- Section 8.3.3 is an interesting aside and speaks specifically to those organizations choosing to outsource asset management, and the requirement to make sure the SAMP is achieved by this third party.

The following excerpts are from ISO 55001:

- Section 4.4 instructs us to make sure the SAMP includes documentation of the role of the asset management system in the achievement of the asset management objectives (recall here that the asset management system might be made up of existing organizational systems; it could be inferred that the normal activities of the business should be in support of the assets).
- Section 5.1 tells us to ensure that the SAMP is compatible with the organizational objectives.
- Section 6.2.1 tells us that the asset management objectives shall be established as a part of the SAMP and be updated regularly.

I think the simple ISO 55000 standard says it best when referencing the Strategic Asset Management Plan:

- Section 2.5.3.4 tells us that the SAMP should be the guide for setting up the asset management objectives and describing how the asset management system will work to ensure the objectives are met.
- Section 3.3.2 informs us that the SAMP specifies how the organizational objectives are converted into the asset management objectives.

The Strategic Asset Management Plan is essentially the Rosetta Stone for translating organizational objectives into asset management objectives. Clearly, the entire organization needs to be engaged in this process. That is why it is applicable to *all stakeholders*.

Just a few closing thoughts on the SAMP. This is a communication document and is likely the actual product an ISO auditor will be looking for and will look at during any certification audit or follow-up activity. If your organization is not working towards a certification, you will still need and want a SAMP. This is an important document.

You might consider a few other elements in this strategic offering:

- An executive summary
- Statement of stakeholder's needs
- Operating context of the assets (internal and external)
- Statement of scope for the asset management system (how all the organization's other systems are going to work together)

This provides a summary of information for the reader. Other information that is more 'operational' and must (read that as mandatory) be included in the SAMP:

- Criteria for decision making
- How the asset is going to be managed over its lifetime
- Asset evaluation methods
- Asset review periods

- Method to address risks and opportunities as they relate to the asset

Figure 6-6 is a reflection of where our conversation has brought us so far.

Figure 6-6 The SAMP as it relates to asset management

Administratively and strategically, your organization should be set now to develop the tactical elements of asset management. These are the asset management plans, or the 'stuff' we actually do.

Asset Management Plans

Asset management plans are the tactically executed activities that an organization performs to match the asset's performance against the organizational objectives. This is actually the care and upkeep of the asset in a production or facilities setting. You can imagine that for the other types of assets (human, financial, etc.), this same sense of matching utilization-to-objective is also being executed. This is a total asset management approach.

The roadmap that your organization chooses to follow to achieve value from the capital or physical asset must be one that considers the life of that asset at your location, and looks at the risks and opportunities that are likely to come during that life cycle usage. The entirety of all the activities must be done with fiscal sensibility.

We're actually closer to a complete asset management plan than you might think as you have gotten this far in the book. I want to discuss the importance of fundamentals for just a few paragraphs. What I hope to achieve during this discussion is an awareness in you, the reader, that for the maintainability and reliability of the physical assets we only have a few things that we have to do and really can do. These are the fundamentals. Our goal is to accomplish these fundamentals, consistently, at the highest level of execution.

The Fundamentals

So what are the fundamentals to maintainability and reliability? Why don't you try to answer, and then I will show you what I came up with. List your answers here:

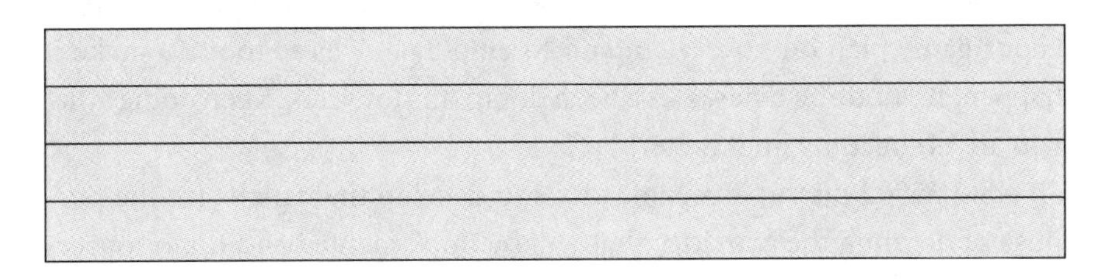

Here is what I came up with:

- Design, build, and install for maintainability (decrease MTTR)
- Design, build, and install for reliability (increase MTBF)
- Asset criticality analysis
- Risk analysis
- An accurate BOM
- Storeroom control and spare parts strategy
- Preventive and predictive maintenance
- Planning and scheduling
- Work order system and control, including work prioritization
- CMMS
- Documentation
- Analysis of maintenance work and reporting
- Data collection and decision-making

- Continuous improvement
- Reliability engineering
- Standardized parts
- Autonomous maintenance (production's involvement)
- Ownership
- Root cause analysis

Go back over the list of fundamentals I provided and match them to the list you created prior to that. Highlight the fundamentals that you and I both listed. How well are you executing these fundamentals?

Our challenge now becomes to use these fundamentals of basic maintenance and reliability and meet the organizational objectives that require the use of physical assets. The company has a strategic intent and part of that intent (cost positioning, beating the competition, product dominance, etc.) might require capital or physical assets. Our asset management plan is our game plan on how we intend to employ our basic tools to make it happen. It's nothing fancy, we've been doing this for years. Keep doing what you are doing, only do it better!

ISO 55002 instructs organizations to develop these plans for the purpose of defining the activities that will be implemented and the resources that will be used to meet the asset management objectives, which we have spent this entire book associating with and linking to the organizational objectives.

My favorite part from ISO 55002 addressing planning to achieve asset management objectives is, "There is no set formula for what should be included or how it should be structured..." (p. 10)

Fortunately for us, that gives us a broad range of possibilities. We know for a fact that our asset management plans need to invoke the fundamental activities mentioned above, and put those into process flows that work to ensure reliability and maintainability of the assets. All of this is happening while we are also maintaining the availability of the assets (the physical assets) to the end goal of meeting the organizational objectives.

ISO 55002 continues to suggest the plan include such details as:

- Rationale for our activities around asset management
- Operational plans

- Maintenance plans
- Risk management plans
- Asset rebuild, replacement, and upgrade plans
- Financial and resource plans

I'd like to break out a few of these ISO 55002 suggested plans in further detail.

OPERATIONAL PLANS

I am listing this first because we need to first understand how operations (production) intends to utilize the asset and how they expect their personnel (operators) to be engaged in the management and upkeep of the asset. For facilities maintenance organizations, give some thought to how the facility users intend to, or need to, utilize the facility assets to their end.

For operational plans, consider the short-term and long-range intentions of the organization for the employment of the asset. How long do they expect to have the asset at your location? Recall from previous discussions that the asset is not useless after your company or plant is done with it. The asset may have value elsewhere.

In the asset management plan, document the operating strategies and how production decisions are made. It is never an operating strategy to run an asset into the ground, although sometimes it seems like that. Keep this point in mind as well; operating costs or the cost of overall ownership is to factor into the total operating plan of an asset.

RISK MANAGEMENT PLANS

What puts the asset at risk or, closer to our purpose, how is the asset at risk of *not delivering* on the organizational objectives? How are risks to assets identified at your location? Provide your answer in the space provided:

I would submit that we put our assets at risk by less than perfect preventive maintenance, poor application of mechanical and electrical skills, shoddy operator care, lack of correct spares, and poor ownership. Do these transgressions exist at your plant or facility and are they identified and dealt with on a timely basis?

And finally, how are risks evaluated and prioritized? Has your organization adopted a risk matrix approach like the one shown in Figure 2-1? In our effort to identify, evaluate, prioritize, and successfully deal with risks, we need to apply the foundational formula that Risk = Probability × Consequence. Some of my clients enhance their proactive approach to risk management by adopting the fundamentals of Reliability Centered Maintenance (RCM) and still others go a step further and incorporate the Failure Modes and Effect Analysis and bring in the principles of detectability, occurrences, and severity to calculate a Risk Priority Number (RPN).

For your own peace of mind and to satisfy any questions regarding risk management, have a plan. Not only document your plan, but have objective evidence that you actually follow your own advice.

MAINTENANCE PLAN

This is us, guys! Our time to shine. This really does mean the maintenance department. Sure, we might fold some autonomous care into our maintenance plans, but it is only for the purpose of augmenting the maintenance department.

Our charter in this sub-section to the asset management plan has to include our process for identifying negative trends, monitoring asset health, and the execution of the fundamental maintenance tasks outlined earlier in this section.

Include in this plan how maintenance tasks are prioritized, maintenance costs, and an explanation on how risk management is based on seeking and providing the best technical advice possible.

Figure 6-7 shows where we are currently on our road to asset management discovery.

I'm submitting to you, the reader, that an entire book can be written covering just asset management plans. In fact, several website postings, magazine articles, and book chapters have been dedicated to this singular

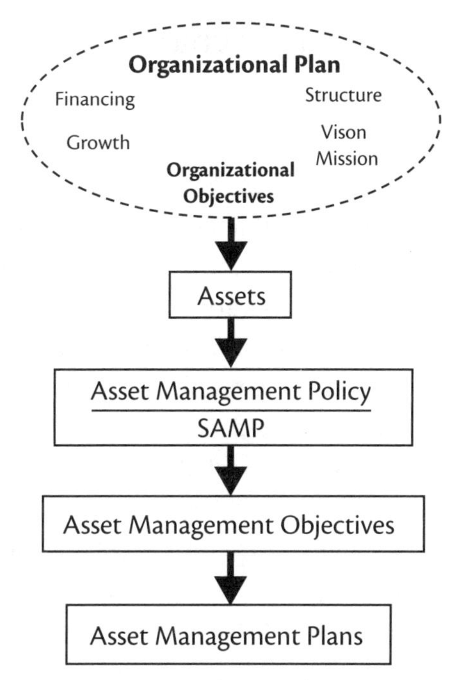

Figure 6-7 Asset management plans in relation to asset management

subject. They all say the same thing: "Here is what ISO 55000 says, and here are our suggestions." My offering is a little more than that.

I strongly suggest that those vying for a formal ISO 55000 Asset Management certification purchase and read the ISO 55000 series of standards, contract the services of a dutiful consultant (or build one internally) and begin to put your plan together. Document the plan, follow the plan, audit the plan, get back on the path of the plan, and succeed because you stuck to the plan.

I'm going to use the remainder of this book to build a short approach to an asset management plan. Again, this is my recommendation, but I want to leave you with a little bit more than just my suggestion. I thought it might be helpful to see what some of this documentation might actually look like.

For our purpose, let's build a scenario around a small midwestern manufacturing company. For my readers who manage facilities or service operations, these steps apply to you as well.

The Scenario

I'm going to reach back and bring in a scenario that I have some practical experience with. I mentioned in the introductory portions of this book and have alluded to the fact that I have worked or consulted in many industries, in fact almost all that anyone can think of, and I have worked or consulted in countries all around the world. But I want to go back to my first civilian job at that pots and pans manufacturing facility. That is where I got my real first introduction to a belief that maintenance personnel were the only ones who cared for and were responsible for asset management. It took time, a consultant, and grit, but we changed all that.

Revere Ware Corporation was a wholly owned subsidiary of Corning, Inc. Revere was founded in 1801 and became the world's premier copper-bottom cookware manufacturer. I tell my classes that Paul Revere himself started Revere Ware since Paul was a copper-smith. I don't know if that is true or not, but it always makes for a fascinating story. It has been years since I worked at Revere Ware, so much of this information is dated, but for our purposes this will work. I have some real information from my time there; other information, for the purpose of this scenario, I will borrow from other locations to make the point I want to drive home.

As the parent company, Corning had the following customer-facing focus (note: this information is taken from Corning's website):

Corning is one of the world's leading innovators in materials science, with a 167-year track record of life-changing inventions. Corning applies its unparalleled expertise in glass science, ceramics science, and optical physics, along with its deep manufacturing and engineering capabilities, to develop category-defining products that transform industries and enhance people's lives. Corning succeeds through sustained investment in RD&E, a unique combination of material and process innovation, and deep, trust-based relationships with customers who are global leaders in their industries.

Corning's capabilities are versatile and synergistic, which allows the company to evolve to meet changing market needs, while also helping our customers capture new opportunities in dynamic industries. Today, Corning's markets include optical communications, mobile consumer electronics, display technology, automotive, and life sciences vessels. Corning's industry-leading products include damage-resistant cover glass for mobile devices; precision glass for advanced displays; optical fiber, wireless technologies, and connectivity solutions for state-of-the-art communications networks; trusted products to accelerate drug discovery and delivery; and clean-air technologies for cars and trucks (http://www.corning.com/worldwide/en/about-us/company-profile.html).

Extrapolating from this information on the website, let's just say for our discussion that Corning has the organizational plan of (in bullet form):

- Maintaining dominance in R&D for glass, optical glass, and ceramic products
- Scaling R&D to full production when it proves profitable
- Capitalizing on market trends
- Helping customers capture new opportunities
- Facilitating trust with customers and communities

Recall from earlier in this chapter that an organizational plan is the critical, necessary, and vital first step to an asset management process. The plan's 'details' are the objectives, and organizational objectives are relevant to the business. For this instance we are talking about Corning as our example. The plan and the objectives would come from the strategic planning accomplished at the highest levels of the corporation.

I'm not privy to the Corning, Inc. organizational plan, but let's say for our purpose that what I've sketched out is close, and those bullets are the organizational objectives. If you read the website description of Corning, Inc., you might agree that this is close enough for this scenario. Again, I want to stress, this is just a scenario I am setting up as an example. These are not Corning, Inc.'s real objectives.

I want to bring back in briefly Figure 2-3 for a short discussion. In Chapter 2, it may have seemed as though I left this figure unfinished, or at least I never referenced it again. I did so with the intent of re-introducing it here for this scenario. Here is Figure 2-3 again.

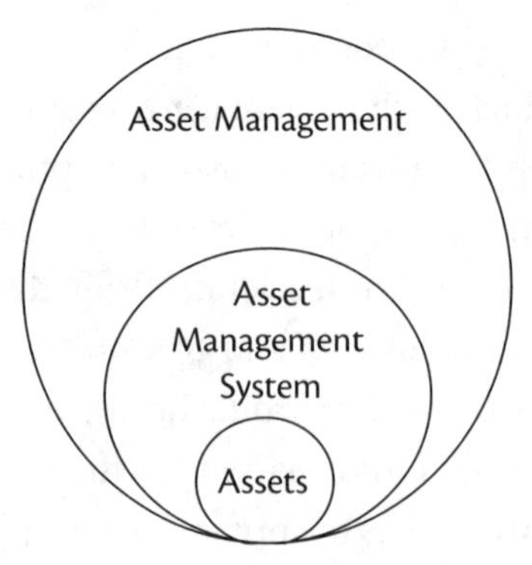

Figure 2-3 Assets within the asset management system (reprinted)

Corning Inc. is a big, big company. Revere Ware was a big company. As a wholly owned subsidiary of Corning, Revere Ware had their own administrative divisions that reported, dotted line, to the giant corporation—independent, yet dotted lined.

Both Corning and Revere Ware have assets they managed. These are all the assets that ISO 55000 references and that we have touched on in this book. Our focus has been on physical assets, as it will be in this scenario as well. What was interesting about Corning that was unusual at the time was that they had, at the highest level, a person who had a chief responsibility to control capital, or physical assets. Corning was a very progressive company back in the 1990s. They still are.

The systems, or the asset management system as shown in the reprint of Figure 2-3, were the cumulative systems of the corporation. Similar to your own company, Corning's HR, safety, engineering, purchasing, procurement, and other divisions all worked in the same software and spoke the same language. It was all in concert to presumably deliver the product to the customer and work toward the myriad of organizational objectives. Those objectives that required physical assets were but a part of a much larger set of corporate objectives. Corning is an extremely responsible company to the communities they operate in, and many of the objectives are set to foster that local care.

When a customer bought a Corning ceramic cooking product, glassware, even an iPhone (the glass on the front is a Corning product), the customer (or stakeholder) was benefiting from this orchestration of systems working together. I mentioned in the ISO 55000 discussion in Chapter 2 that the stakeholders include the customer, employees, and others. I think that Steve Jobs himself was instrumental in working with Corning on the iPhone glass R&D.

That's a good review, but I am driving to a point. It is the little circle in the Venn Diagram of Figure 2-3 that I want to bring into play. When I arrived at Revere Ware in 1994, it looked like it must have looked when Paul Revere left it (that may just be rumor). There was no structure, organization, or maintenance and reliability process. Just good people fighting the good fight. I literally started at the beginning.

Many reading this book know my personal journey. It was documented in my first book. To keep on this scenario, I want to fast-forward to the point that we (the Revere Ware leadership) had to determine what assets (the small circle) we were going to focus on. This is really where our scenario starts.

If you have not highlighted a single entry in this book, highlight this one. It is vital that we carefully select what assets to include. ISO 55000 gives the organization the flexibility and the room to identify exactly what assets are going to be covered as part of the formal asset management effort. ISO 55002 states, "The boundaries and applicability of the asset management system should be captured in a statement of scope." (p. 5)

Asset Criticality

Revere Ware Corporation had 5,600 individually identified assets. Since they made pots and pans, much of the equipment was big, old, heavy, and very dirty. There were several multi-hundred-ton presses, acid plating tanks, material handling equipment, knurling machines, rolling machines, stamping machines, etc. I think you can get a sense of the types of assets we had. Nothing was new. Everything was old and worn out. I had an air compressor that was literally off of a sunken WWII battleship.

Our first task was to perform a criticality analysis to determine which of our assets really mattered, and which ones mattered the most. I do not have any record of the criticality analysis that we performed. To be frank, we didn't even have computers back then at our little plant. At least I didn't have much more than a desktop with a green screen, Word Perfect, and a dot-matrix printer.

As a substitute, let me show you a RIME chart (Ranking Index of Maintenance Expenditures), which has an asset criticality ranking off in the left margin. I will bring this RIME chart in for further discussion later. Figure 6-8 is the RIME chart detailing the asset criticality in the left margin.

	Rank	10	7	6	5	2	1
Rank	**Equipment**	**Breakdown Critical Safety**	**PMs**	**Corrective**	**Service**	**Routine**	**Facilities and Grounds**
10	Main Power Air Compressor IT Network Furnace	100	70	60	50	20	10
9	Main Water Cutting Table 6 Line 1 Line 2	90	63	54	45	18	9
8	Cutting Table 8 Line 8 Line 00	80	56	48	40	16	8
7	Line 3 Line 7 TTUM 2000 STM	70	49	42	35	14	7
6	BND 17 Cutting Table 4/5 MUTT 4000 Hawkeye	60	42	36	30	12	6
5	Batch Saw Filler TFB Bender Sling Crane	50	35	30	25	10	5
4	Pattern Area Chopper Spider	40	28	24	20	8	4
3	Forklift Sweeper Docks	30	21	18	15	10	3
2	Trash compactor Rec compactor	20	14	12	10	4	2
1	HVAC	10	7	6	5	2	1

Legend
Priority 1
Priority 2
Priority 3

Figure 6-8 RIME chart

I want to be clear on Figure 6-8 that this is an example of a RIME chart and not the results of our criticality analysis of assets at Revere Ware. I did not have Excel back in 1994. I probably recorded our criticality analysis and filed it as a paper document.

Look on the left margin at the ranking and 'weight' of the assets of this particular plant. That is the criticality of the plant assets. This ranking is arrived at by forming a small team of stakeholders (production, engineering, maintenance, leadership) and just hashing it out. I offer this cautionary note to you the readers. I would recommend reserving the top spot, the uber-critical asset space, for listing only utilities. I know production assets are critical, but nothing is more critical than getting electricity into the facility.

Once we had completed something similar to the criticality ranking achieved while building a RIME chart, we had the assets listed in order of importance, or criticality, and we had an idea which ones we needed to focus on and focus on first. But now what?

Using the Risk Matrix

I am going to reprint Figure 2-1, the risk matrix introduced in Chapter 2.

At Revere Ware we created such a document. Our risk matrix did not look much like the one in the reprint of Figure 2-1. To be honest, we didn't really know what we were doing. But, to be fair, we did have really good intent. Here is how you would read and use such a matrix. Keep in mind that Risk = Probability (frequency) × Consequences (severity).

We had several hundred-ton presses. In fact, they were multi-hundred-ton presses. Reading the risk matrix, here are the questions I asked:

- If we performed no maintenance at all, no service, no inspections, and no 'fixes,' what is the frequency of failures we could expect to see?
- If that failure happened, what would be the severity?

Figure 6-9 would be a typical response to the questions I just asked for these hydraulic presses.

SEVERITY / FREQUENCY	FREQUENT (A) ≥1 per 1,000 Hours	PROBABLE (B) ≥1 per 10,000 Hours	OCCASIONAL (C) ≥1 per 100,000 Hours	REMOTE (D) ≥1 per 1,000,000 Hours	IMPROBABLE (E) <1 per 1,000,000 Hours
CATASTROPHIC (I) Death or permanent disability Significant environmental breach Damage >$1M, downtime >2 days Destruction of system/equipment	HIGH	HIGH	HIGH	MED	ACCEPT
CRITICAL (II) Personal injury Damage >$100K and <$1M Loss of availability > 24 hours and <7 days	HIGH	HIGH	MED	LOW	ACCEPT
MARGINAL (III) Damage >$10K and <$100K Loss of availability >4 hours and <24 hours	MED	MED	LOW	ACCEPT	ACCEPT
MINOR (IV) Damage <$10K Loss of availability < 4 hours	ACCEPT	ACCEPT	ACCEPT	ACCEPT	ACCEPT

Figure 2-1 A risk matrix (reprinted)

SEVERITY / FREQUENCY	FREQUENT (A) ≥1 per 1,000 Hours	PROBABLE (B) ≥1 per 10,000 Hours	OCCASIONAL (C) ≥1 per 100,000 Hours	REMOTE (D) ≥1 per 1,000,000 Hours	IMPROBABLE (E) <1 per 1,000,000 Hours
CATASTROPHIC (I) Death or permanent disability Significant environmental breach Damage >$1M, downtime >2 days Destruction of system/equipment	HIGH	HIGH	HIGH	MED	ACCEPT
CRITICAL (II) Personal injury Damage >$100K and <$1M Loss of availability > 24 hours and <7 days	HIGH	HIGH	MED	LOW	ACCEPT
MARGINAL (III) Damage >$10K and <$100K Loss of availability >4 hours and <24 hours	MED	MED	LOW	ACCEPT	ACCEPT
MINOR (IV) Damage <$10K Loss of availability < 4 hours	ACCEPT	ACCEPT	ACCEPT	ACCEPT	ACCEPT

Figure 6-9 A typical response to a risk matrix study

Adapted with permission by Marshall Institute, Inc.

This was no surprise, really. I mentioned that our equipment was old, and it had been abused since Paul Revere's time. Just to put this conversation in perspective, we are developing the strategies around our assets and their functions in our operating plants, within the operating context of our facilities. This is all done with the presumption that Corning and Revere Ware needed physical assets to meet their organizational objectives.

Using the positing shown in Figure 6-9, our job in the maintenance department for this particular asset became to move that 'circle' down and to the right. In a risk matrix, you want to move the target from the position of where it is currently, down and to the right. We do this by performing top-notch and world-class maintenance. The fundamentals.

As an example, Figure 6-10 indicates where I would personally like to move the risk (frequency × severity).

Using the example shown in Figure 6-10, our task became figuring out what to do in order to drive the risk from the previous spot shown in Figure 6-9 to the location shown in Figure 6-10.

I'll cut to the chase on this. Here is what we did:

- Created a preventive maintenance program around that asset
- Scheduled and executed those PMs
- Found corrective work from the PMs that were conducted
- Scheduled the repair of that corrective work
- Created a BOM on that asset
- Stocked parts in the storeroom based on the BOM
- Responded to unscheduled work on that asset based on the RIME chart's priority setting
- Rinsed and repeated

How much of this have we already covered in this book? Quite a bit. Everything that I just listed had a formally documented process around it. When I say documented, I actually mean printed in Word Perfect. As stated earlier, our computing and storage capacity was not strong back in the late twentieth century.

Before I get too far off the subject of the risk matrix, let me summarize its use and value.

FREQUENCY / SEVERITY	FREQUENT (A) ≥1 per 1,000 Hours	PROBABLE (B) ≥1 per 10,000 Hours	OCCASIONAL (C) ≥1 per 100,000 Hours	REMOTE (D) ≥1 per 1,000,000 Hours	IMPROBABLE (E) <1 per 1,000,000 Hours
CATASTROPHIC (I) Death or permanent disability Significant environmental breach Damage >$1M, downtime >2 days Destruction of system/equipment	HIGH	HIGH	HIGH	MED	ACCEPT
CRITICAL (II) Personal injury Damage >$100K and <$1M Loss of availability > 24 hours and <7 days	HIGH	HIGH	MED	LOW	ACCEPT
MARGINAL (III) Damage >$10K and <$100K Loss of availability >4 hours and <24 hours	MED	MED	LOW	ACCEPT	ACCEPT
MINOR (IV) Damage <$10K Loss of availability < 4 hours	ACCEPT	ACCEPT	ACCEPT	ACCEPT	ACCEPT

Figure 6-10 The aim

Adapted with permission by Marshall Institute, Inc.

A company (mine and yours) needs to employ physical assets to reach organizational objectives. It is the responsibility of everyone in the organization to work toward the availability of the asset to perform its needed function. We use the risk matrix to compel others as well as ourselves to do 'something' to reduce the risk of the asset failing and not adding value. Once the risk matrix shows us that we must do something, then the organization determines what the acceptable level of risk (like in Figure 6-10) is for our requirements. Do you recall that I asked you in the beginning of this book what "acceptable level of risk" meant to you?

How do we lower the risk? It is simple. The maintenance department performs awesome maintenance. The production department keeps the equipment clean, operates it properly, and escalates maintenance as needed for issues. Procurement gets us the right parts. HR performs the onboarding and maintains the training matrix and gap analysis. Everyone is pulling in the same direction toward the organizational objectives that were born from the organizational plan.

There are some fundamentals left for us to discuss.

The RIME Chart

I mentioned that I was going to bring the RIME chart back in for discussion. Rather than reprint it, please refer back to Figure 6-8 for this brief discussion.

You have already noted the asset criticality listed down the left margin. Note that along the top are the maintenance activity codes, listed according to weight or severity. The RIME chart is a series of columns and rows. Using the weight of each, every individual cell becomes the multiple of that row and column's weight. For example, a cell with a value of 63 is the product of a column with a weight of 7 and row with a weight of 9; $7 \times 9 = 63$. If you look at Figure 6-8 once more, for this client you will see that 63 is in a light gray cell and the legend to the right tells us that this is a priority 2.

From working with this client, I can tell you that for a priority 2, their maintenance will commence in 7–14 days on that assignment. The RIME chart is the tool for the decision-making criteria. Do you recall that decision-making criteria is a requirement of ISO 55000?

The risk matrix is also decision-making criteria, as is the Stock, Don't Stock Decision Tree. You have many examples of decision-making criteria in this book.

Let's discuss three more of these fundamentals: preventive maintenance, planning and scheduling, and organizational structure.

Preventive Maintenance

I have always been curious about knowing whether to perform a PM or PdM (predictive maintenance). If you ever wondered why we call it preventive maintenance when it seems that we are actually performing maintenance, it's because we are preventing 'greater' and more involved (expensive) maintenance. For example, we keep our car tires inflated to the proper amount regularly (small maintenance) to keep from replacing the tires more often (greater, more expensive maintenance). We are essentially *preventing* more maintenance.

For the decision of performing preventive or predictive maintenance, I would recommend a RCM logic tree or what some may call a PM/PdM Decision Tree. In either case, it is also an example of decision-making criteria. See Figure 6-11 for an example of a PM/PdM Decision Tree.

Look at Figure 6-11 and recognize that our first attempt to eliminate a failure potential is to redesign the asset, or engineer the issue 'out.' This would be our action for a really egregious concern. If that is not possible, our next strategy is to devise a predictive approach.

Why do you think predictive maintenance is preferable over preventive maintenance for the permanent address of failures or failure modes? See if you can think of any reasons and record those here:

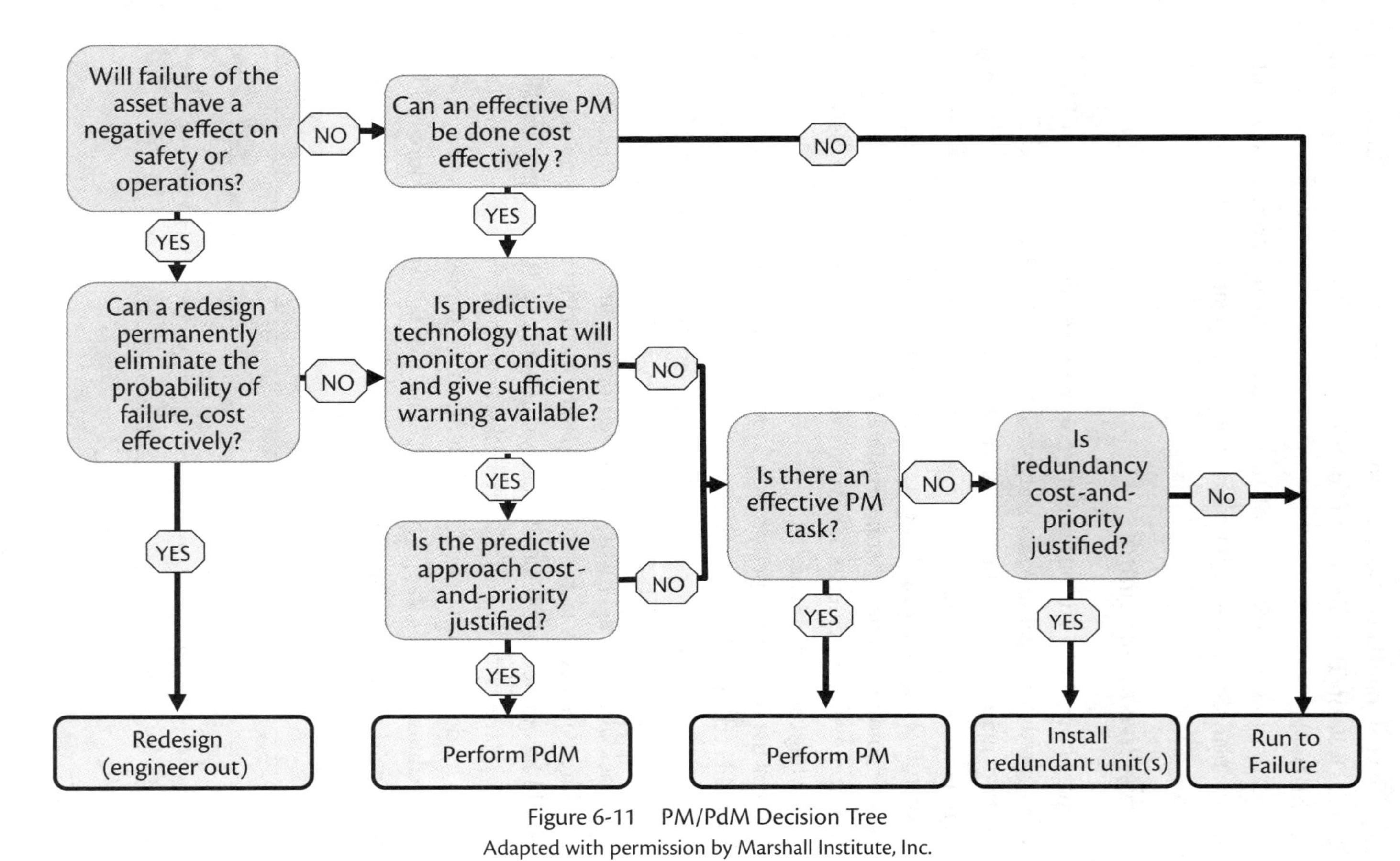

Figure 6-11 PM/PdM Decision Tree
Adapted with permission by Marshall Institute, Inc.

There are three principle reasons that I can think of and these should be all the ammunition you will need to compel leadership to finance a predictive maintenance effort:

- The result is always a dot on a chart that can start a trend to follow.
- The result is always an objectively agreed upon value (everyone can see the thermometer reading).
- The predictive action is taken while the machine is running and producing product.

A note about preventive maintenance. I strongly suggest adopting a methodology such as Preventive Maintenance Optimization (PMO) to scrub and enhance your PMs and bring them back into relevance. What follows is the before and after results of a PMO workshop. PMO did not exist when I worked at Revere Ware, and we used another process that was not nearly as robust.

Look at this example of the power of PMO and notice that in each instance, the new PM task is instructing the technician on:

- What to do
- How to do it
- What 'good' looks like
- What to do if it's bad
- Any safety steps to be aware of

BEFORE PMO

This is an honest to goodness real semi-annual PM.

Time: 16:11:51
User: USDOTH

WO no	0006567069	
Project no		
Priority	3	PRODUCTION
Order type	Z55	PREVENTIVE REQ
Criticality cl	3	PRODUCTION
Document ID	RUSSELL SIEVE SEMI ANUAL PM	
Ref order no		
Event type	OPMO999924	SEMI ANNUAL PM
Equipment no	Z11.SC592	RUSSELL EUROPA 1200 SIEVE
NH position	Z11.OPMZM402A	PA1_SCREEN/SIEV
Production line	OPMSCR01	SCREENING

Requested by		Tel no
Approved by		Tel no
Reason code	CODE	
production stop	0	
Start date	092515	
Finish date	092515	

LUBRICATE TOP BEARING 4 GRAMS
LUBRICATE BOTTOM BEARING 4 GRAMS
CHECK BAND CLAMPS FOR TIGHTNESS
CHECK ALL HOSES & BELLOWS FOR CRACKS OR LEAKAGE
CHECK FOR LOOSE FASTENERS
CHECK CONDITION OF MOUNTING SUPPORTS ON FLOOR
CHECK PIPING FOR LEAKS
INSPECT MOTOR COUPLING
INSPECT WAGON WHEEL MOUNTS

SEMI ANNUAL PM RUSSELL SIEVE

Op description	RUSSELL EUROPA 1200 SIEVE		
Work center	ZOOGPRE	0010	MAINT. PM WORKS (STONY CREEK)
Employee no			
001			

MOS143WO-A4

Notice that there are two bearings on this machine that require very little greasing. All these inspections and service steps are performed semi-annually. This is a vibrating machine that sifts a very aggressive product.

What follows is the after PMO workshop PM package. Some of you are not familiar with a process known as PM by exception. Don't be fooled by monthly PMs being followed directly by quarterly, and so on. This is a unique process that the client needed. Do note, however, that the new PMs are of a higher caliber and the PM is written to reflect exactly what is needed and expected. For my PM writing style, I always include the storeroom part number for the component addressed in the PM, if applicable. Also, at the end of the PM, I list the sources for the data.

POST PMO PM

Here is the actual new PM product.

WORK ORDER

WO no		
Project no		
Priority	3	PRODUCTION
Order type	Z55	PREVENTIVE REQ
Criticality cl	3	PRODUCTION
Document ID	RUSSELL SIEVE SEMI ANNUAL PM	
Ref order no		
Event type	OPMO999924	PM
Equipment no	Z11.SC592	RUSSELL EUROPA 1200 SIEVE
NH position	Z11.OPMZM402A	PA1_SCREEN/SIEV
Production line	OPMSCR01	SCREENING

Requested by		Tel no
Approved by		Tel no
Reason code	CODE	
Production stop	0	
Start date	092515	
Finish date	092515	

MONTHLY PM RUSSELL SIEVE

Skill: Technician

Time Est: 15 mins

Tools/Equipment:

- Calibrated grease gun (2 grams/stroke)
- Mobile XHP222
- Clean rags

Vibrator Assembly

1. Top Bearing

While machine is running, with a clean rag, wipe grease zerk clean, apply 4 grams of Mobile XHP222 grease using calibrated grease gun, with a clean rag, wipe grease zerk clean (bearing part #ZO00002407).

OK	ADJ	CM

2. Bottom Bearings

While machine is running, with a clean rag, wipe grease zerk clean, apply 4 grams of Mobile XHP222 grease using calibrated grease gun, with a clean rag, wipe grease zerk clean (upper bearing part # ZO00002407; lower bearing part #ZO00002402).

OK	ADJ	CM

Figure 1 Typical Russell Sieve bearing grease zerks

QUARTERLY PM RUSSELL SIEVE

SAFETY: APPLY LO/TO
Skill: Technician
Time Est: 30 mins
Tools/Equipment:

- Flashlight
- Metric and Standard wrenches
- Allen wrenches
- Thin scraper
- Clean rags

Vibrator Assembly

3. Top Grease Line

Visually inspect grease line to top bearing looking for signs of crimping, cracking, or other evidence of line compromise. There should be none. If line is damaged notify maintenance lead. If line is not attached properly to top casting and grease zerk at base assembly, reattach.

OK	ADJ	CM

4. Bottom Grease Line

Visually inspect grease line to top bearings looking for signs of crimping, cracking, or other evidence of line compromise. There should be none. If line is damaged notify maintenance lead. If line is not attached properly to top casting and grease zerk at base assembly, reattach.

OK	ADJ	CM

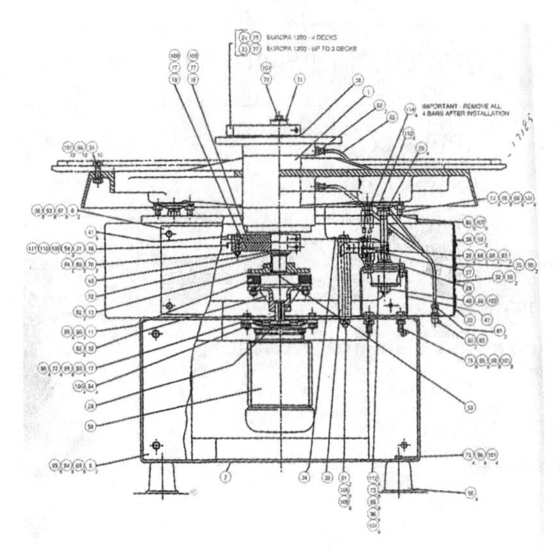

Figure 2 Grease line schematic

5. Flexible Coupling

Visually inspect flexible coupling for evidence of excessive normal wear, dry rot, and missing material; there should be none. If evidence exists, notify maintenance lead. (Flexible coupling part # ZO00109010).

OK	ADJ	CM

6. Flexible Coupling

Using a wrench, confirm that the flexible coupling fasteners are tightly secured to the drive flange. Additionally, ensure the set screws on the drive and the driven coupling flange are secure.

OK	ADJ	CM

7. Flexible Coupling

Using a scraper device, or can of compressed air, clean the gap between the flexible coupling and the driven flange. There should be a gap between these two surfaces.

OK	ADJ	CM

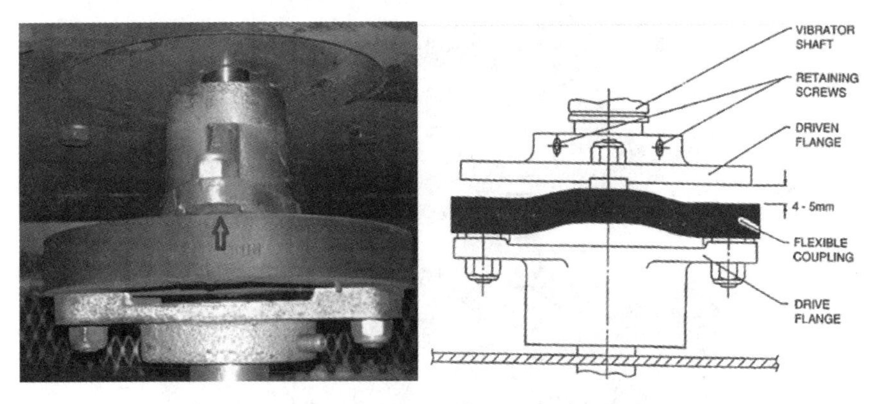

Figure 3 Flexible coupling and schematic

8. Wagon Wheel

Visually inspect wagon wheel mounts (4x top and bottom) for evidence of wear, cracking, tearing; there should be none. If evidence exists, notify maintenance lead.(Top and bottom suspension mounts part #ZO00109050)

OK	ADJ	CM

9. Wagon Wheel

Visually inspect wagon wheel tie rods (4) for evidence of mounts cracking or bending, there should be none. If evidence exists, notify maintenance lead. (Suspension rod [tie rod] part #ZO00001223)

OK	ADJ	CM

10. Wagon Wheel

Using a wrench, confirm that the fasteners mounts are tightly secured.

OK	ADJ	CM

Figure 4 Wagon wheel mounts

11. Motor Using a wrench, confirm that the motor mount fasteners are tightly secured.

OK	ADJ	CM

SEMI ANNUAL PM RUSSELL SIEVE

SAFETY: APPLY LO/TO
Skill: Technician
Time Est: 20 mins
Tools/Equipment:

- Calibrated grease gun (2 grams/stroke)
- Polyrex EM grease
- Metric and Standard wrenches
- Flashlight

- Clean rags

Vibrator Assembly

12. Motor

With a clean rag, wipe grease zerks clean, apply 4 grams (0.16 ozs) of Polyrex EM grease using a calibrated grease gun, with a clean rag, wipe grease zerks clean. [**IF APPLICABLE**]

OK	ADJ	CM

Figure 5 2-HP motor

13. Weights

Visually inspect weight (2) settings, each should be at position G; if the weights are not at position G, notify the applicable process engineer.

OK	ADJ	CM

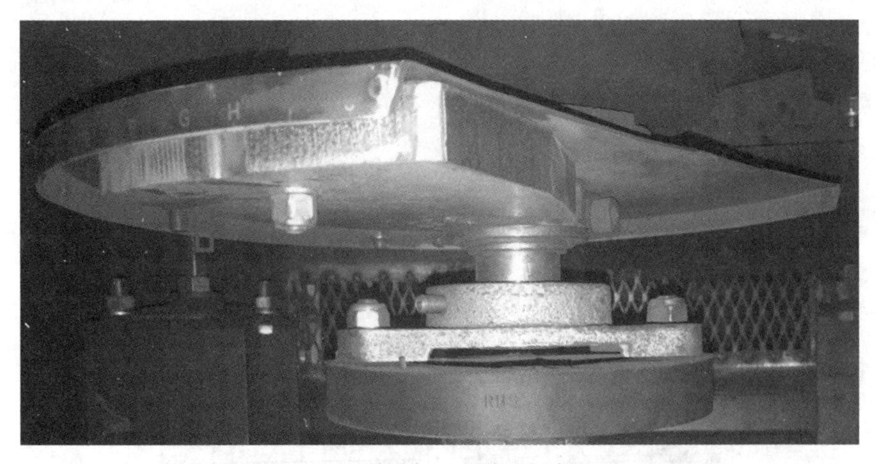

Figure 6 Weight position indicator

14. Weights

Using a wrench, confirm that the weight fasteners are tightly secured; if loose, tighten.

OK	ADJ	CM

Figure 7 Weight fasteners and hub set screw

COMMENTS (Please document adjustments made and corrective maintenance requirements):

____

____

____

____

____

____

____

____

____

____

____

____

Technician Signature and Date

Source Material:
RUSSELL EUROPA 1200 SIEVE Owner's Manual
RUSSELL SIEVE SEMI ANNUAL PM (only PM document prior to 2/11/16)
Maintenance Lead
Maintenance Engineer
Area Process Engineer
Area Operator
Electrician
Sr. Process Engineer

If you read through this new PM you might have noticed that there are actually three bearings, and the owner's manual instructs the user to grease each bearing while the machine is running, on a monthly basis. Did you also note that many components were not even addressed on the old PM? Of special interest is the removal of the requirement to check the hoses on this machine. There are no hoses on this machine.

Your asset management plan needs to include notice that it is your intention to drive to this level of preventive maintenance and predictive maintenance, if warranted and necessary.

The resulting corrective work from preventive and predictive maintenance is the number one input to maintenance planning and scheduling.

Planning and Scheduling

Planning and scheduling are two entirely different roles. Planning maintenance work is concerned with the 'what' and the 'how.' Scheduling work is the 'when' and on the day of execution, the 'who.'

Planning activities should center around *what* is required to be accomplished, and exactly *how* to go about doing it. Scheduling, as just mentioned, is *when* it needs to be done, and *who* will be assigned to accomplish the job. They are two different roles with many responsibilities, but often performed by the same person.

I addressed scheduling in the previous chapter and gave you a formula for calculating a run time that would be conducive to an effective run reliability of >85%. I would strongly urge any production leader to consider using this formula to develop at least an educated guess of when an asset (based on history) is 'needing' service and maintenance. Don't ignore your own data. Remember, ISO 55000 requires a data-driven effort.

I would like to touch on the planning function and suggest that every maintenance organization have a dedicated maintenance planner who is left to plan maintenance. It sounds odd to write it that way, but I usually encounter maintenance planners who do everything but actually plan maintenance.

A planner should be at the same organizational level as the maintenance supervisor. With the unique job titles that exist in plants and facilities today, let me clarify that by saying that the planner and the maintenance supervisor should report directly to the same person.

A planner *will* (I want to stress that) only be concerned with approved work orders. Each work order will be *walked-down,* meaning that the planner will visit the job site. While at the site, the planner should use a planning checklist to document what is needed to complete the work. The checklist might include such information as:

- Work order number
- Asset number
- Priority number
- Tools needed
- Equipment needed
- Material needed
- Parts needed
- Major procedural steps to complete the job
- Safety precautions
  - Energy isolation
  - Electronic interlocks
- Other outstanding work orders for the same asset
- Estimated time to complete the work order
- Skills required and number of people required for the work order

What other information could be collected by 'walking-down' a job?

That information will be used to complete the maintenance job plan for that particular assignment. The job plan will contain information such as listed here:

- Prints
- Calibration details
- Parts list
- Photos
- Copies made from the technical manuals
- Feedback form (to communicate changes with the planner)

What other items would you put in a job plan?

The value of a well-written job plan is not exhausted after the job, not by a long shot. As we created job plans at Revere Ware, we filed the documentation to be reused at another time. It is a guarantee that if you change out a particular pump, motor, hydraulic unit, etc., you will be changing out that same component sometime in the future. Now my organization had a pre-developed *plan* of how to do just that.

The secret to filing and reusing job plans is in creating the governance for how you are going to label and retrieve these documents. I hate to admit it, but back when all our files were paper files it was easy to find the right document. Electronic filing of job plans can be a real challenge when arriving at a naming convention and storage strategy.

Figure 6-12 is an interesting and amazing picture of the job plan and work order paper filing system of one of my clients. Imagine figuring out how to set this all up in the CMMS and then being able to find it ten years later.

Figure 6-12 Job plan and work order filing system
Adapted with permission by Marshall Institute, Inc.

Does Figure 6-12 look like the files at a doctor's office to you? That's ok, it did to me as well. The information contained in that room is the collective knowledge of the history of that 150-year-old manufacturing facility. They documented everything. What you see in Figure 6-12 is the older information. Since this company adopted a CMMS they have been saving their files in the CMMS. I can assure you that their computer filing system is every bit as accessible and user-friendly as their paper filing system was.

Just because it's fun, I thought I would share the notice that is pinned to the wood paneling right above the chief planner's head in this very same facility. Figure 6-13 is meant to be fun, yet poignant.

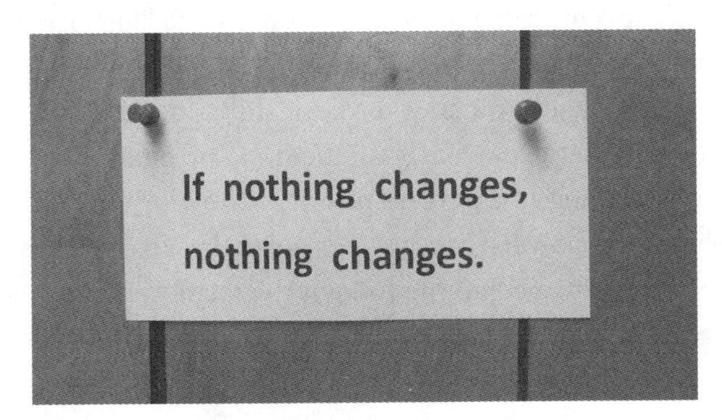

Figure 6-13 A fun reminder

Organizational Structure

I'm going to touch briefly on the subject of the organizational structure of the maintenance group. You may recall that earlier in the book I referenced a magazine article I wrote years ago regarding my belief that we (maintenance) set ourselves up to be reactive simply by the way we are structured. I used that reference in this text to suggest that the same problem exists in the production department. I have covered a lot of ground since then, and I thought I would share with you where I was coming from and suggest a better way forward. This is in alignment with ISO 55000's requirement to be resourced to be successful.

When I arrived at Revere Ware Corporation in 1994 there was no supervision at all. There was no planner, scheduler, preventive or predictive maintenance. There was no CMMS, no work orders, not even a written work order system. There was literally no storeroom at all, not even a junk closet. Through the help of a consultant from Marshall Institute (recall the story earlier in this book where I said that I worked with one consultant and that I called him in) we moved from 100% reactive to 98% uptime in eighteen months.

Figure 6-14 is the structure I left Revere Ware with after several years of building a great proactive program.

What made the structure depicted in Figure 6-14 so 'proactive?' Of course, you will notice that the planner and the shift supervisors are all on the same organizational level. Note that we actually had a planner. We also hired a reliability engineer and created a storeroom. The storeroom lead-person (Kathy, an hourly union associate) performed exceptionally well, just like Paula my calibration lab person (same plant). They were both very bright and capable ladies and a pleasure to work with.

Notice that we had a 'hit crew' on each shift. Their job was to respond to breakdowns as they occurred, but if there was no emergency at the time, they were functioning to complete priority 3 (low priority) work orders. The day shift had two very proactive groups. The first was the PM crew, who wrote and performed all the PMs in the facility. The second was the line mechanics who performed all the corrective work.

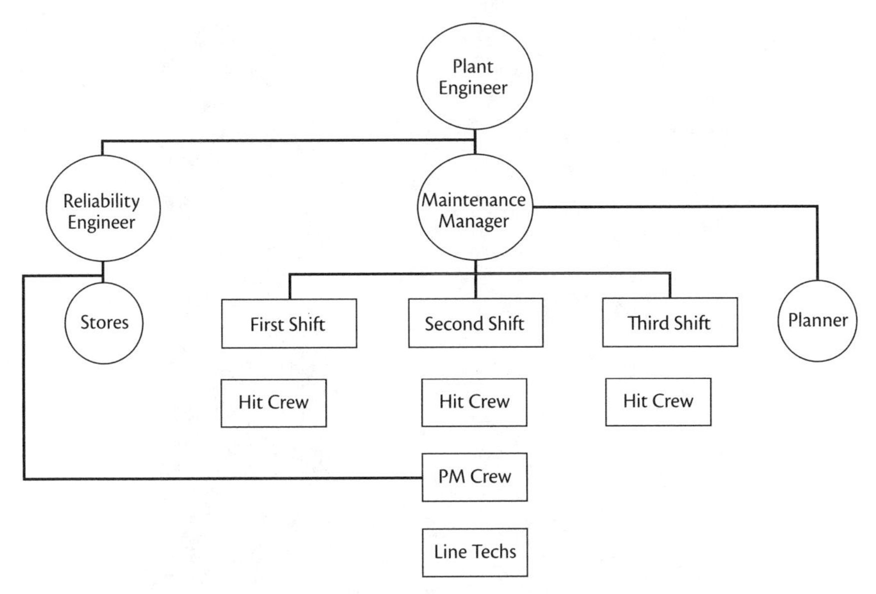

Figure 6-14 A proactive maintenance structure

I would hope that someone reading this book would ask, "How can you say you had a proactive structure when you had three crews dedicated to breakdowns?" That is a great question.

What made this a proactive organization was that I was set up to find issues (PM crew) and fix issues (line mechanics) while I was also keeping the emergency work at bay. Isn't it true in your organization that the emergency work keeps pulling folks off of the planned and scheduled work?

Look, we are always going to have emergency work to do. Let's face it, as long as people are driving fork trucks in our buildings they are going to run into something and create an urgent situation. Very few plant bollards are still standing straight up; almost all are listing at 11°. Does Figure 6-15 look familiar?

Figure 6-15 Do your plant bollards look like this?

The consultant I hired, Jack, told me that maintenance needs to perform three things every time, and at the same time, and they need to do them all well: preventive maintenance, corrective maintenance, and emergency maintenance. My first civilian boss, Mike, would say, "Always give yourself a chance to be successful." I learned both of those lessons at the same place: Revere Ware Corporation.

I hope you learned some tips from that scenario. My intention was that you might gain a sense of what objective evidence might look like for some of the activities that you and your team must develop and execute. Chapter 6 has been a long chapter, but I want to leave you with one final thought, and it is an overview of an asset's life. I think you will find this to be interesting and possibly helpful.

An Asset's Life

This is an overview of an asset's life. I've done some research on this matter and I was hoping to tie in an old story that I shared with you toward the front of this book. Please recall the story I told about my friend Terry at Wurtsmith AFB in Michigan. Remember that Terry worked at the motor pool and had told me that the Air Force gets rid of its staff cars after they have expended the balance of their 'account.' Keep that story in mind as you read through this last section.

I believe we must use the idea of the estimated life cycle cost of an asset to help aid in overall asset management. This brings in the final elements of the asset management system shown initially in Figure 2-6. Let me explain.

Ramesh Gulati has an interesting table in his book *Maintenance and Reliability Best Practices*. I have reproduced some of that information here in Table 6-1. (p. 177)

Table 6-1 Asset Life Cycle Cost (LCC) Breakdown-For Industry

Cycle	Estimated Costs (%)
Design and Development	5–10
Installation	10–20
Operations & Maintenance	65–85
Disposal	<5

The numbers, according to Gulati, vary slightly for DoD (Department of Defense) life cycle costing. Table 6-2 shows the numbers my friend Terry faced.

Table 6-2 Asset Life Cycle Cost (LCC) Breakdown—DoD

Cycle	Estimated Costs (%)
Design and Development	10–20
Installation	20–30
Operations & Maintenance	50–70
Disposal	<5

The Air Force chose this approach when it came to purchasing, using, and disposing of staff cars. At least they seemed to follow this approach at my base. See Figure 6-16.

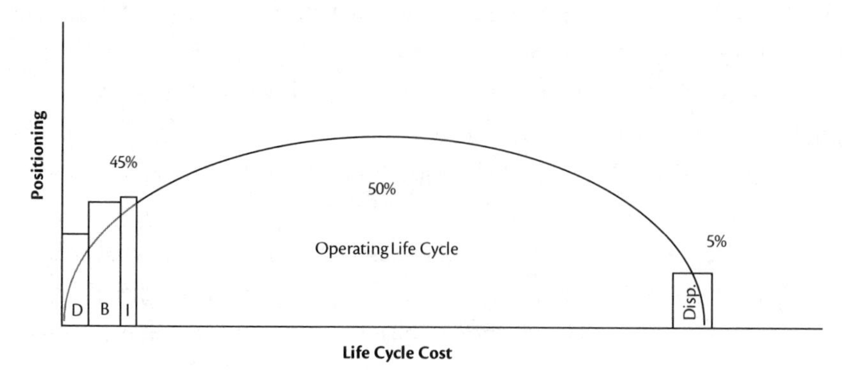

Figure 6-16 The life cycle of an Air Force staff car

Table 6-3 reflects the numbers Terry communicated to me.

Table 6-3 Life Cycle Cost of a Staff Car

Cycle	Estimated Costs (%)	Costs ($)
Delivered	45	8,600
Operations & Maintenance	50	9,555
Disposal	5	955
	Total	19,110

According to Table 6-3, and following this logic, the Air Force had decided long before they fielded a staff car at Wurtsmith Air Force Base in the 1980s that the most they would spend on a single car was $19,110. That is asset management! How can we use all this information?

I'm glad you asked. Let's look at Gulati's data graphically in Figure 6-17, which is aligned with Table 6-1.

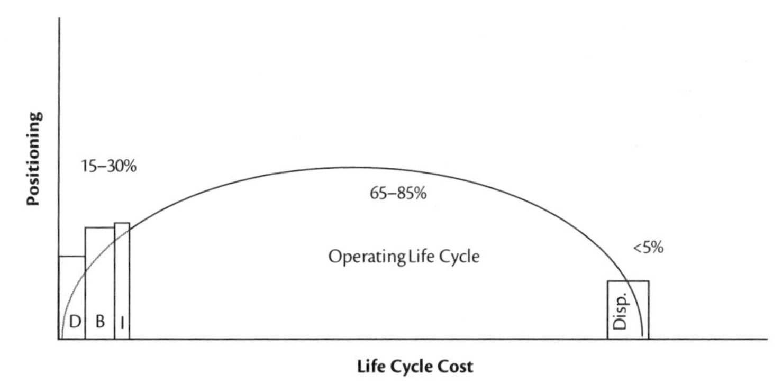

Figure 6-17 Industrial Life Cycle Cost spread

Let's pull in a scenario where we use the graphic in Figure 6-17 to make a larger point. As an example, we have an interest in purchasing a new conveyor system that is an off-the-shelf system and has a sticker price, installed, of $250,000. Using some information we have just gained in the closing section of this chapter and throughout this book, what approach should we take to ensure this asset adds value and can be used as part of our effort to reach organizational objectives?

We have an opportunity to depreciate this conveyor system over a ten-year period and we anticipate that the conveyor system will have a value of $50,000 at the end of that ten years. Our depreciation calculation becomes (straight line depreciation):

> $250,000 – $50,000 = $200,000
>
> $200,000/10 years = $20,000 per year. That is what our annual depreciation expense for the machine would be instead of an upfront $250,000 expense that would be a significant hit to our company's net income.

For an asset life of ten years at our facility, Table 6-4 is the range we could expect to see for Life Cycle Cost using the graphical information shown in Figure 6-17.

Table 6-4 LCC Range for Our $250,000 Conveyor Asset

Design, Development, and Installment	Operations & Maintenance	Disposal	LCC Total
15%	80%	5%	100%
$ 250,000	$ 1,333,333	$ 83,333	$ 1,666,666
20%	75%	5%	100%
$ 250,000	$ 937,500	$ 62,500	$ 1,250,000
25%	70%	5%	100%
$ 250,000	$ 700,000	$ 50,000	$ 1,000,000
30%	65%	5%	100%
$ 50,000	$ 541,666	$ 41,666	$ 833,332

Using Table 6-4 as a guide, if we utilize our decision-making criteria and determine that the second installment where O&M costs reflect 75% of our LCC is our preferred approach to asset acquisition, then we have some information that we can use to lay out the next ten years of experience with that conveyor system in our facility.

Figure 6-18 now becomes what we are working with.

Our company has made an asset management decision indicating that we are going to purchase a conveyor system for $250,000 and keep it for ten years. During that time, we are expecting to spend $937,500, or $93,750 per year on O&M costs. At the end of ten years, the conveyor will have a value of $50,000 on our books, and we have set aside $50,000 to remove it and dispose of it. Hopefully, we can sell it to another facility so it can continue to add value to another company.

We can manage this budget by performing the asset management steps that have been laid out in this book: good solid fundamental maintenance and reliability practices that engage all the stakeholders. By recognizing this asset as one that helps to create the paycheck for all our associates, we should be motivated to do what we need to do to keep this asset as a viable value winner for our company.

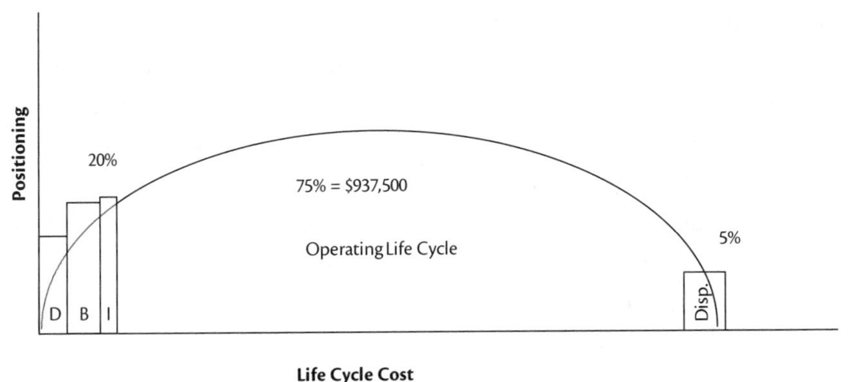

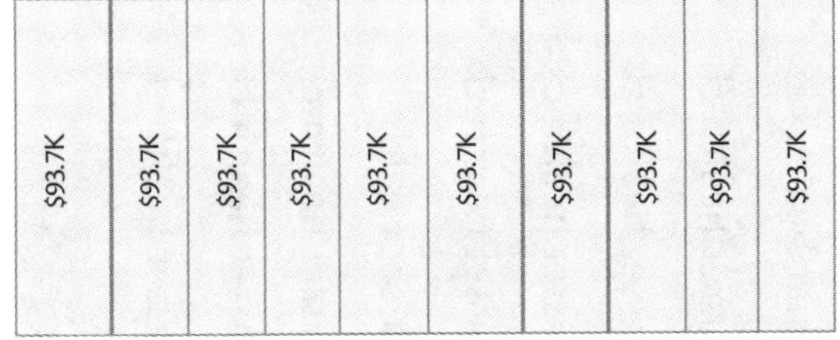

Figure 6-18 LCC costing for the conveyor system

We will set up the BOM and arrange for spare parts as we have learned to do. We will encourage through team building and training that production fill their positions as asset managers. Production will be engaged in the escalation of maintenance if there are any problems occurring on this conveyor system outside of normal and routine preventive and predictive maintenance.

We will further assess this asset each year to determine the health of the asset through a review of the work orders, parts replacements, and overall performance and function toward the operational objectives, which stem from the organization's production goals. This evaluation will be paired with improvement opportunities in an effort to ensure its life in our plant is successful and productive.

I had a student break down their approach to asset health by telling me that they bring in their predictive maintenance contractors periodically for scheduled 'sit-downs' with the plant personnel to work out a reasonably accurate understanding of the current health of the asset at that moment in time. They have a checklist that a combined (plant-contractor) team runs through on an asset and the findings are presented in a traditional red/green display. With that they are able to definitively say what they believe the life time is left on the asset. I have to say, I was rather impressed. In deference to the student, I have only shared what I felt comfortable sharing with you on their processes.

With that particular example in your mind, review Figure 6-19 for an idea of how the asset's life and performance against budget can be constantly assessed, managed, and improved upon over the course of the asset's time in our facility. I've include the PM/PdM/CMs and yes, Emergency Maintenance (EM) for each year—these are the fundamentals!

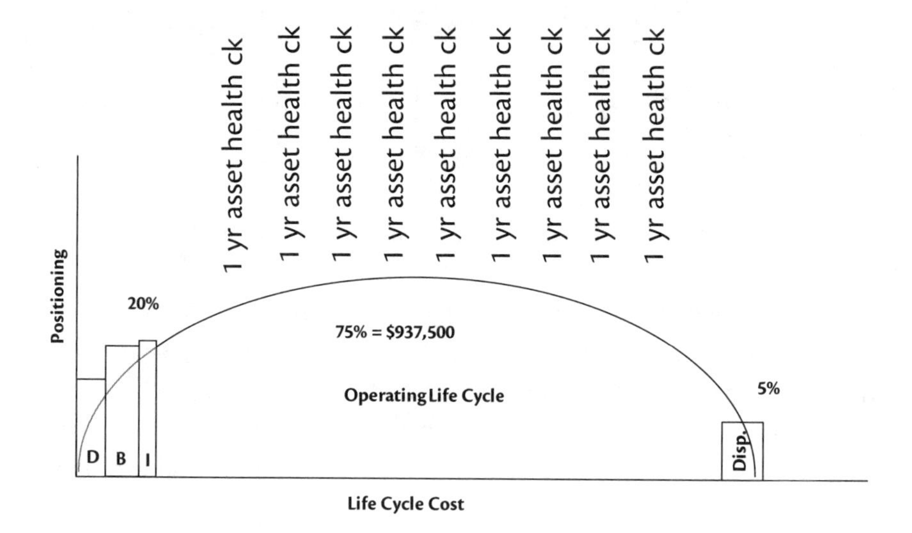

$93.7K	$93.7K	$93.7K	$93.7K	$93.7K	$93.7K	$93.7K	$93.7K	$93.7K	$93.7K
PMs	PMs	PMs	PMs	PMs	PMs	PMs	PMs	PMs	PMs
PdM	PdM	PdM	PdM	PdM	PdM	PdM	PdM	PdM	PdM
CM	CM	CM	CM	CM	CM	CM	CM	CM	CM
EM	EM	EM	EM	EM	EM	EM	EM	EM	EM

Figure 6-19 The asset's constant health checks

I am looking back on Figure 6-19 and thinking about the maintenance on my own car and the maintenance on my own health. Think about yours as well. Doesn't our approach to the asset (our car or our health) sort of look like Figure 6-19? It may even include a maximum we are going to spend for the upkeep. On our car, not our health!

Well, just like that, we have arrived at the end. Figure 6-20 shows us where our discussion in Chapter 6 has led us.

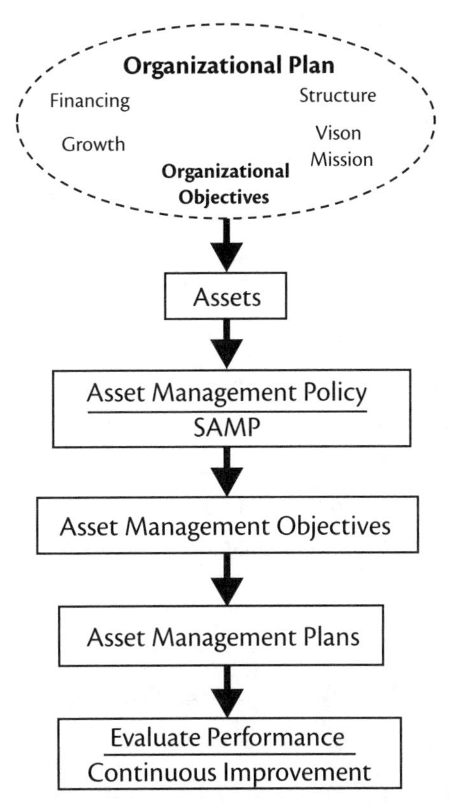

Figure 6-20 The asset management system

As I promised, I had fun writing this book. I hope that you had fun reading it and that you found it to be informative and helpful. Please let me hear from you if you felt it was useful and were able to apply what you've learned. This has been a learning experience for me as well.

CHAPTER SUMMARY

Asset management at your place and at your pace will take the combined effort and interest of all stakeholders. Fortunately, the ISO 55000 standard calls for just that level of activity. As we discovered in this chapter, there is enough work to go around. However, and this is important, nothing can start as far as formally aligning asset management with the corporate directions until the organizational objectives are determined. These objectives are determined by the overall organizational plan.

Adding to Our Business Case

The organizational plan is the corporate plan. The organizational plan is the critical, necessary, and vital first step to an asset management process. An organizational plan includes great detail about how the company intends to be successful. These 'details' are the objectives, and organizational objectives are relevant to what we are discussing. These objectives come from the strategic planning accomplished at the highest levels of the corporation.

The plan should be communicated in such a way that all parties are knowledgeable as to what our organization is all about and the impact it hopes to have in the space in which it operates.

Our company's organizational plans are:

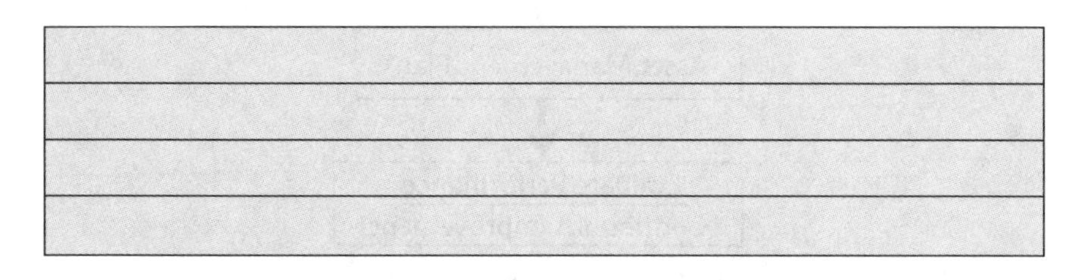

The plan provides the details on the objectives. These objectives, as well as others, help to define the trajectory of the organization and effectively point the direction the company must follow to ensure that all its assets (physical, human, financial, etc.) are working together (in harmony) to reach the goals of the organization.

The mission statement that guides our company should include a description of what our organization holds as its core values. Our core values and the objective evidence that we hold to these values include:

Before we can defeat our enemies we have to name them. The chief competitor in our space is:

The organizational objectives should be far-reaching and lofty, yet speak to the common man. ISO 55000 is intending for our organization to take the conversation of these core objectives down to the floor level.

Our company's organizational objectives are:

Similar to any policy, an asset management policy is a statement of intent. This intent should address the guiding principles that are important to the organization and be supportive of the organizational plan and the organizational objectives.

For our asset management policy, we will consider evidence that clearly indicates that:

- The policy was created by stakeholders
- The policy is followed

- The policy is reviewed from time to time to ensure it still supports the organizational objectives

The SAMP is the transformational document that converts organizational objectives to asset management objectives. A well-conceived Strategic Asset Management Plan will also lay out the approach to be taken to comply with the overall asset management objectives.

The SAMP is an iterative process and is forever being evaluated and improved. An iterative process is a process that gets ever closer to the desired results the more iterations that are conducted, similar to the Deming Wheel.

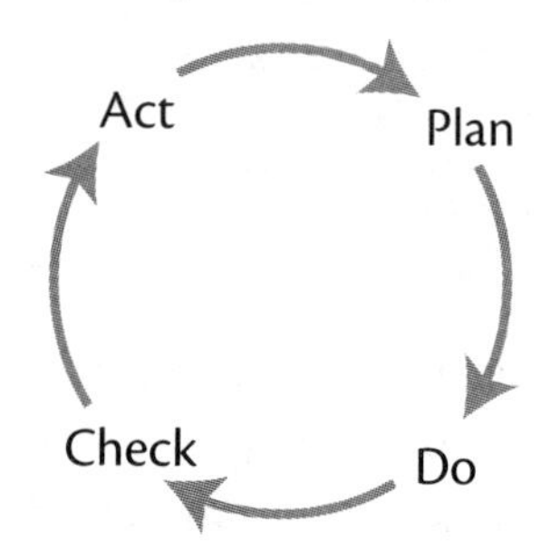

Our organization needs to show that there is constant reviewing and aligning of the SAMP and asset management objectives with the organizational objectives. These reviews become the objective evidence that the SAMP is a living product and is 'freshly' in tune with the direction of the company.

Our asset management objectives include:

The SAMP will be the guide for setting up the asset management objectives and describing how the asset management system will work to ensure the objectives are met.

Asset management plans are the tactically executed activities that an organization performs to match the asset's performance against the organizational objectives. This is actually the care and upkeep of the asset in a production or facilities setting. The activities that we are involved in for asset management are the fundamentals of good maintenance and reliability listed here:

Our goal is to accomplish these fundamentals, consistently, at the highest level of execution.

ISO 55002 instructs organizations to develop these asset management plans for the purpose of defining the activities that will be implemented and the resources that will be used to meet the asset management objectives. ISO 55002 continues to suggest the plan include such detail as:

- Rationale for our activities around asset management
- Operational plans
- Maintenance plans
- Risk management plans
- Asset rebuild, replacement, and upgrade plans
- Financial and resource plans

The final stages of an effective asset management system include the measure of asset performance against the project value to meeting organizational objectives. A constant fine-tuning is required through a robust continuous improvement environment.

Summary

I've been involved in asset management my whole professional life. Of course we didn't call it that back then, but that is what it was. I don't mean just physical asset management; I have been engaged in all of it. I suspect many of you have as well.

Although I was primarily focused on the actual plant equipment, I got involved in financial assets, personnel assets, IT assets, property assets, etc. I was asset management! I never once had the warm fuzzy feeling that other division chiefs were as engaged in physical asset management as I was engaged in their asset management systems. To me, asset management, or physical asset management, was just a maintenance effort. This belief may permeate through your organization today as well.

We need to change that. It is the very equipment that is working in your plant or facility today that is helping to generate the paychecks of all the employees. Not only that, but by effectively using the assets, the company makes income. Physical asset management affects everyone and therefore requires the attention of all stakeholders.

I opened this book with a catchy phrase, "I don't think you're smart enough." You know, I was wrong. Not only are we smart, but we are capable. We don't have to overcome the laws of physics to be really good at asset management. We just need to execute the fundamentals consistently at the highest level.

And finally, we need to up our documentation game. We need to say what we do and do it.

Thank you for sharing this experience with me. Now, go and do something awesome!

The Master Business Case

We often begin our journey in asset management by setting off on the wrong foot. The manifestation of that misstep is in our acceptable definition of key words and phrases. For the purposes of our discussion on asset management, here are some key phrases and the working definitions at this location:

Maintainability:
Reliability:
Asset:
Asset Management:

It must be the focus of this organization to recognize that there are many types of assets, including but not limited to:

- Capital assets
- Human Resource assets
- Financial assets
- Property assets
- Trade secrets or proprietary assets
- Processes
- Inventory
- Accounts Receivable
- The company's brand

For the extended purpose of asset management, we are going to focus on capital assets, recognizing that the effective deployment and utilization of capital assets make income for the company. Assets, it turns out, are what we manipulate to make income. Robert Kiyosaki animated this for us in the following figure.

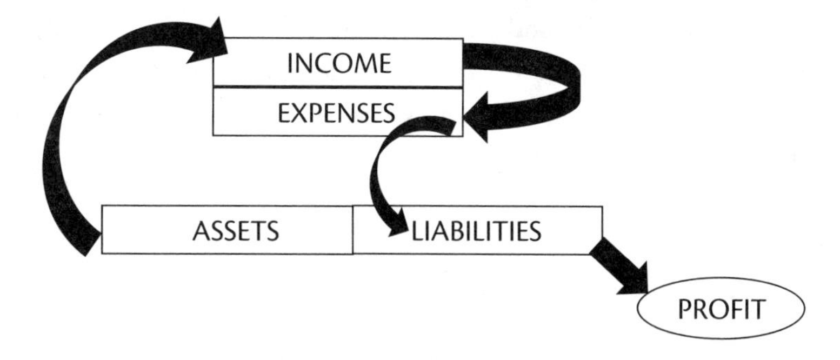

In each case, in order to extract the most income, these assets must be honored and cared for as if the balance of the organization was hanging from their success. Because it is.

Our focus is on capital assets. In a way, all of the members of our company have to see the connection between their work and the proper maintenance and reliability of the company assets (capital equipment). It is with this connection that we all become engaged in *asset management*.

Through this connection to the asset and its purpose of creating income for the enterprise, we have to articulate exactly how this happens. The following is an example of how a single plant asset is manipulated to make income for the company.

List one capital asset:	
How is it manipulated for financial gain, means #1:	
How is it manipulated for financial gain, means #2:	

Further, to demonstrate the importance of the company's assets to the business aim of this organization, here at this location, our care manifests itself as shown in this example:

List one capital asset:	
Demonstration of asset's importance #1:	
Demonstration of asset's importance #2:	

It is strongly believed and virtually dictated in the ISO 55000 that the employees be given a voice in equipment design and construction. As a result, we suggest that all design teams have a standard makeup to include the following representation:

- Engineer
- Maintenance leadership
- Production leadership
- Maintenance hourly
- Production hourly
- Safety
- Purchasing
- Storeroom

These teams require the structure of a widely accepted methodology to adopt continuity and direction. Our organization will follow ISO 55000 Asset Management.

For asset management we choose to adopt the suggested and often mandated principles set forth in the ISO 55000 series of documents: ISO 55000, ISO 55001, and ISO 55002. As such, our intended audience resides in one or more of the following areas:

- All those engaged in determining how to improve the returned value for the company from their asset base (we are primarily focused on physical assets)
- All those who create, execute, maintain, and improve an asset management system
- All those who plan, design, implement, and review the activities involved with asset management

The primary benefit to adopting such a methodology is to ensure a sustainable approach to meeting organizational objectives with the performance of the organization's assets. We intend to adopt a risk-based approach to asset decision making. What follows is an example of an asset risk matrix.

SEVERITY / FREQUENCY	FREQUENT (A) ≥1 per 1,000 Hours	PROBABLE (B) ≥1 per 10,000 Hours	OCCASIONAL (C) ≥1 per 100,000 Hours	REMOTE (D) ≥1 per 1,000,000 Hours	IMPROBABLE (E) <1 per 1,000,000 Hours
CATASTROPHIC (I) Death or permanent disability Significant environmental breach Damage >$1M, downtime >2 days Destruction of system/equipment	HIGH	HIGH	HIGH	MED	ACCEPT
CRITICAL (II) Personal injury Damage >$100K and <$1M Loss of availability > 24 hours and <7 days	HIGH	HIGH	MED	LOW	ACCEPT
MARGINAL (III) Damage >$10K and <$100K Loss of availability >4 hours and <24 hours	MED	MED	LOW	ACCEPT	ACCEPT
MINOR (IV) Damage <$10K Loss of availability < 4 hours	ACCEPT	ACCEPT	ACCEPT	ACCEPT	ACCEPT

It is through the management of the risk, and the capture of opportunities that our organization can realize value in their assets and have a desired balance of cost-to-risk-to-performance.

The primary task of the asset management approach is to ensure organizational assets work to achieve the organizational objectives as applicable. Our understanding of the organizational objectives is:

We will continue to show that the assets are performing in alignment with the organizational objectives by including the following metrics in our report on value-to-performance:

In addition to becoming an organization that reaches risk-based conclusions, we will also be an organization that is data-driven and follows the decision-making criteria listed here:

Our associates will be competent and empowered as evidenced by our practice of:

There are external and internal forces that make up the operating context of our facility. Those may include, but are not limited to:

EXTERNAL CONTEXT

- Social
- Cultural
- Economic
- Physical environments
- Regulatory
- Financial

INTERNAL CONTEXT

- Organizational culture
- Environment
- Mission
- Vision
- Organizational values

Our approach to establishing an asset management culture will be in line with the following Venn diagram:

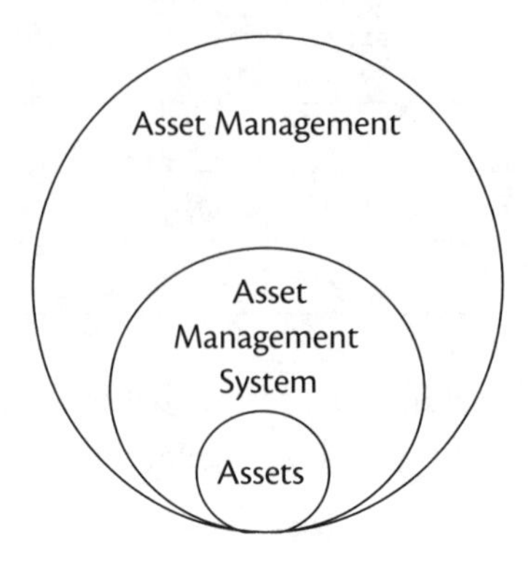

The organization determines exactly what is and what is not included in the asset management system. To be certain, as it relates to physical assets or capital assets, a *scope* can be introduced in the SAMP (Strategic Asset Management Plan) to address these boundaries.

These boundaries are to be communicated to the internal and external stakeholders. For our purposes, we have determined that the following assets will be under the auspices of the formal program:

Our asset management activities will contain contingency planning for when events occur, and opportunity seeking efforts when those instances arise.

We are required to have the proper resources to achieve the organizational objectives relevant to asset value, and those resources have to be:

- Competent
- Aware
- Communicated with
- Informed

It is fiscally responsible and good common sense to build an asset management system from the systems that are already present in the organization. Those systems are (e.g. work order system, planning and scheduling system, hiring system, accounting system)

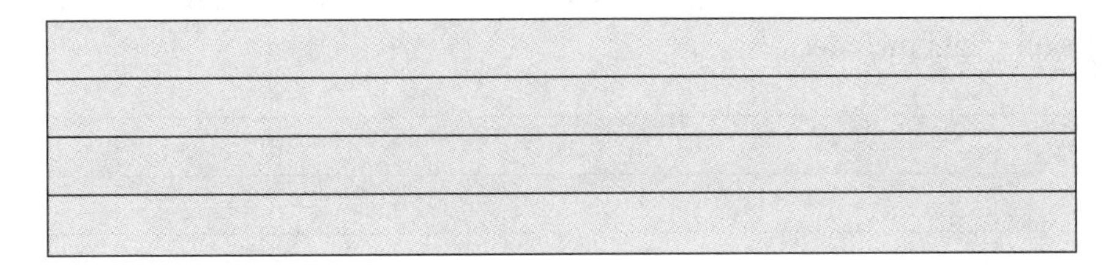

What follows is a representation of the major elements to our asset management system:

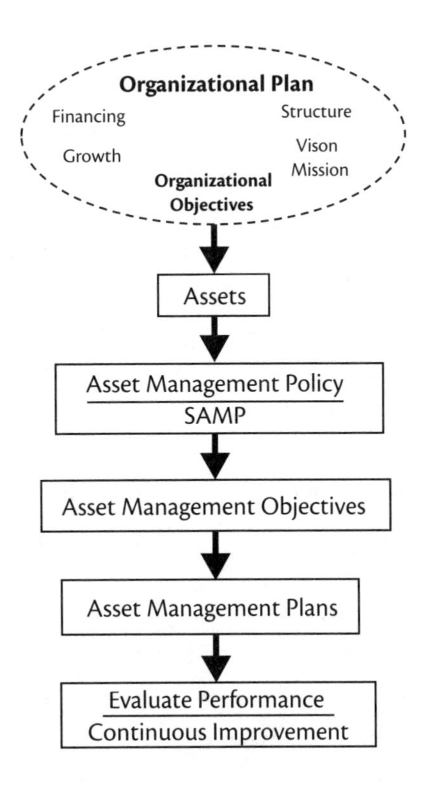

To function within this asset management system, we need data to make asset decisions. These decisions center on whether it is in the best interest and in alignment with organizational objectives to purchase a new asset or modify an existing asset. This is in concert with the ISO 55000 requirement to be data-based, and have decision-making criteria.

An analogy to this thought is one where consideration is given to purchasing a new car. An example of the data needed to make a new car decision might include:

When purchasing a new asset or modifying an existing machine, maintainability and reliability should be initial elements for consideration and overall requirements.

ISO 55002 instructs a qualifying organization to have, as its plan, the rationale to perform the value-preserving activities that it (the organization) is mandating. Furthermore, the plan is to include maintenance and operational necessities, and there must be measurements in place to determine if the plan is working. The activities involved in improving maintainability and reliability are the 'preserving activities' referenced.

The rationale to decrease MTTR by increasing maintainability is an increase to asset availability. If done right, this single element could also improve product output and overall productivity.

Maintainability is synonymous with MTTR and is generally measured in four distinct elements:

- Response time, meaning:

- Troubleshooting time, meaning:

- Repair time, meaning:

- Startup time, meaning:

To decrease an asset's MTTR, we need to shorten one or several of these four elements. As an example, we can do this by:

The rationale to improve reliability is also an increase to asset availability. Reliability is synonymous with MTBF and is commonly considered the time between an asset failing, or breaking down, and the asset being brought back up into production. Not only does the asset have a MTBF value, but the components on the assets have their individual MTBF values as well.

As an example, our current asset:

has a MTBF. That asset is made up of several components, five of which are:

These five individual components each have their own MTBF. To increase an asset's MTBF we need to understand what components are contributing to the asset's reduced MTBF.

One of the chief ways to understand each component is to adopt a philosophy of parts standardization. Parts standardization contributes to the strategic asset management plan by:

As mentioned, an important necessity for improving MTBF is to identify the MTBF of the asset, and then break that down and see what the MTBF-limiting components are. Our job is to know these components and know which components are responsible for limiting the MTBF. From there, small teams will study the data and devise plans for improving the Mean Time Between Failure. This is a very simple, continuous process.

MTBF and MTTR matter to our business because these two reliability metrics make up the Inherent Availability measure as shown in the following formula:

$$A_i = MTBF/(MTBF + MTTR)$$

The further we drive MTTR to zero the closer we get to A_i = MTBF. That would be an ideal state.

After assets are purchased and installed or modified the work is not done. If not completed prior to the asset startup, PMs need to be modified or created, and a comprehensive listing of spare parts needing to be stocked will require the storeroom's engagement. Spare parts and the preventive maintenance protocol on assets square with the mandates of the asset management plan and the SAMP. These actions are the 'activities' referred to in the ISO standards. They are literally the 'stuff we do.'

Preventive maintenance is another aspect of operating a new asset that is every bit as critical as the spare parts. Although each asset in our plants is generally made up of the same components, each of those components is running in a different environmental and operational context.

The spare parts strategy and the preventive maintenance strategy for each asset in our facility is part of the inherent makeup of the asset management plan. These are integral elements of the activities that we actually do. In fact, preserving the inherent reliability of an asset is contingent upon the practice of like-for-like parts replacement and executing preventive maintenance steps as prescribed.

Thinking strategically, we have to adopt the mandated requirement that we draft PMs prior to starting the machines in our production effort, and that we have on hand the necessary operational spares and a strategy to keep both current and relevant.

We recognize that access to the correct spare parts, when needed and in the correct quantity, contributes to the overall health and reliability of the assets. Furthermore, since assets arrive at our facility with inherent reliability, it is the role of everyone to ensure that the inherent reliability is maintained. One way to accomplish this is to have a supply chain guaranteeing all the spare parts when needed.

The ISO 55000 series of documents is riddled with pronouncements that the organization must have systems and processes in place to measure, evaluate and essentially deal with risks of all kinds. Who can deny that the all-consuming, time-intensive process of locating and procuring the necessary spare parts is putting our organization at risk?

One necessary element, buried in ISO 55002 to be specific, is a section on management of change, with a small note that the organization should address supply chain constraints as a means to avoid an increased leveraged risk by this particular choke point (that being the lack of access to spare parts). This is rationale enough to address, at the highest level, what most of us have known for a very long time. Namely, that spare parts or the dearth of them are our Achilles heel!

Using a stock, don't stock methodology, we will determine what components to stock in our on-site storeroom. We will only stock items for an asset that appear on that asset's Bill of Materials, or BOM. A BOM is a list of all the individual components that make up the parent component. For each new or modified asset, we will demand an as-built BOM, provided in

English, in writing and electronically. These BOMs will only be accepted for the 'as-built' asset.

A BOM is often depicted in an exploded-view diagram such as this:

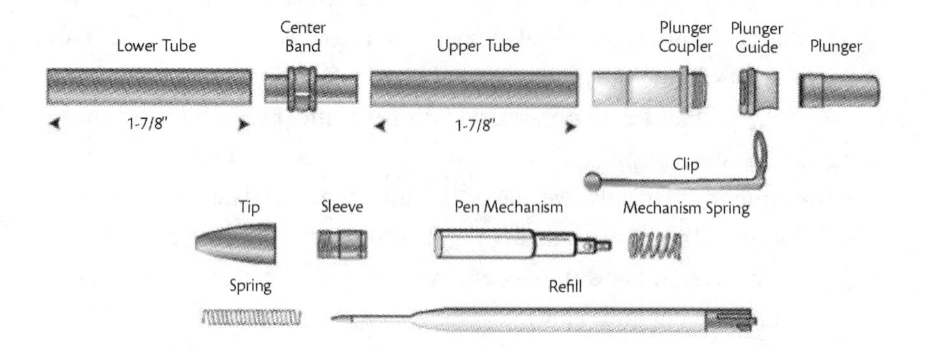

The diagram is accompanied by a listing of all the components on the asset, such as this:

1.	10.
2.	11.
3.	12.
4.	13.
5.	14.
6.	15.
7.	16.
8.	17.
9.	18.

Our organization will retain the as-built BOMs for each asset and the associated parts listing. Further, these parts will be arranged into a parts hierarchy similar to the following:

The project engineer assigned to the capital project to purchase a new or modify an existing asset will have the primary responsibility to deliver the complete BOM, thereby meeting the criteria recently defined. The planner has the responsibility to oversee the BOMs with updates and changes as the asset is utilized and maintained while in the performance of its function in our facility.

A proper and complete BOM has a positive and lasting effect on almost all plant agencies. Following is an example of the agencies that benefit from a Bill of Materials:

Production:
Engineering:
Storeroom:

Procurement:
Maintenance Management:

For those components that appear on an asset's Bill of Materials, and we determine should be stocked in the storeroom, we will evaluate and optimize the stocking method. For sporadically used parts, the best method to use is the traditional min/max level. We will apply a concept known as safety stock where appropriate. Beyond min/max, we hope to identify those components that have a stable usage pattern. For those we will apply the Economic Order Quantity (EOQ) approach using the following formula:

$$\sqrt{\frac{2NP}{IU}}$$

- 2 is a constant
- N = annual usage of the item, full count
- P = cost of a purchase order; most folks use an average of $65 per P.O.
- I = carrying costs, percentage as a decimal
- U = unit cost in dollars

Where it makes sense in terms of preserving the asset and its functions for the value of our stakeholders and our shareholders, we will work with our vendors and suppliers to establish Vendor Managed Inventory and/or consignment stock.

In all our storeroom and spare parts efforts, we will formalize the twenty-eight world-class storeroom processes and resource our stockroom so the level of our MRO capability matches what is needed to meet the organizational goals. In each process there will be clearly drawn roles and associated responsibilities.

It isn't just the storeroom that plays a significant outlier role in asset management. To be successful at engaging stakeholders, to include employees, we must fold in the largest group of employees in our organization—production. This starts at the beginning of their tenure.

It seems that during the onboarding process an employee is likely to get more instruction on how to fill out a 'vacation request form' than they do on how to fill out a 'work request form.' That needs to change. Odds are that an associate will need to request work from maintenance more often than they will vacation days from their boss.

This must be a cautionary tale however; the answer is never as simple as showing operators how to fill out a work notification or work request. An operator must first have a sense of the theory of operation of their equipment before they can then determine that something is wrong with it.

At our facility, operators are provided theory of operation training by:

Our asset management policy includes a notice that operators and other production affiliated personnel will have a robust onboarding process that includes an extended theory of operation. The onboarding process for production associates should include an unambiguous exchange of the new employee's responsibilities as they relate to the organization objectives, their responsibilities toward the care and operation of their assigned asset(s), and a very comprehensive plan to measure and assess their knowledge and development of the training protocol to close the gap between their current understanding and the company's expectations.

To exercise that knowledge, operators and production personnel will become an integral part of the overall maintenance and reliability effort through a formal and very well defined escalation of maintenance protocol.

Currently, the steps at our facility to communicate equipment issues flow like this:

Our efforts to hone the skill of certain operators to a higher level of capability and asset care will be enhanced with the addition of an operator-tech position allowing for more focused delivery of the skilled trades where needed. Inputting the role of an operator-tech, the maintenance escalation process would now progress through several in-line skilled and trained operational human assets before diverting a maintenance human resource from their planned and scheduled work:

- From the Operator to his or her Supervisor
- From the Supervisor to the Operator-Tech
- From the Operator-Tech to the Maintenance Technician

Production's role in an asset management world takes on a new set of requirements and stretches the employees into a role that they have been traditionally not engaged in; namely the conception of asset related policies and procedures. Additionally, the operator group must actively participate in the development of measures and performance indicators to determine if the activities surrounding an asset are in line with the organizational objectives. This is a significant role addition and, as such, an equally significant effort will be taken to ensure that our operator base is educated and equipped with the knowledge and skills to be major contributors to the discussion.

For their role in asset-related policies and procedures, production personnel will be engaged in formal training which will be documented on a skills matrix similar to the one shown next:

	Rockwell Hardness					
Operator Requirements Matrix–3/20/19 **LEGEND** **Level 4** Fully capable of performing on own and training others **Level 3** Fully capable of performing on own **Level 2** Can perform with coaching **Level 1** In training	Explain Theory of Operation	Set Up Rockwell Tester	Calibrate Rockwell Tester	Execute Hardness Testing	Complete Test Certification	Troubleshoot Tester
Bill Dunbar						
Randy Auld						

The increased role in the asset's health doesn't stop at this intermediate position. Highly trained and capable operators can now be considered ideal candidates to fill the greater need in the skilled trades through a formally devised and executed apprenticeship program. Selection of successful candidates is paramount to the health of the organization and the continued delivery of asset performance. Imagine a core of exceptionally capable trades people who have 'come up' through the operator ranks to bring a well-rounded wealth of operational and technical knowledge to future asset discussions.

Although the lines of separation between maintenance and production begin to blur and gray around a high performing asset management company, each skill has its place and its time. With that in mind, production will assume the necessary role of pulling maintenance into production to be in line with the synergies of lean manufacturing. It is necessary for production to establish the level of care they wish for the company assets to deliver to ensure the needed performance to match the organization's objectives. Maintenance cannot be forced on production and each side must determine the appropriate time for scheduled maintenance.

The time estimate for scheduled maintenance in our facility will be derived from asset performance history and utilize the reliability formula shown here:

$$e^{-\lambda(t)}$$

- e is the natural log, it is a constant value, which is approximately 2.71828
- λ is the Greek letter lambda, in this formula it represents the inverse of the MTBF
- (t) is the time that the asset is intended to run, based on the same units as the MTBF (hours, minutes, etc.)

Maintenance will provide the best technical advice and it will always center on a system reliability value of >85% using the formula just shown. As an example of this formula in use, consider this scenario:

An A-1 packaging machine is a stand-alone asset and is used sparingly during off-peak production and more often during peak production. You and your team are confident in your calculation of the MTBF of this equipment, and the number you have arrived at is 560 hours.

The MTBF is 560 hours, what is the inverse, or lambda?

The production team wants to run this packaging machine for thirty straight days, around the clock. This is the peak period.

What is the resulting reliability of that scenario?

The continuous run time that would be permissible to have a resulting reliability (probability) of getting through that run time of 85% is:

Using this scenario, the resulting run schedule as advised by maintenance would look like this schedule shown:

90 hrs. running	1 hr. maint.	90 hrs. running	1.5 hrs. maint.	90 hrs. running	1 hr. maint.	90 hrs. running	2 hrs. maint.	90 hrs. running

The engagement of the production department in the full attainment of asset management at our location is crucial, as is the involvement of all stakeholders. In the end, we here must make asset management work at our place and at our pace. To that end, we must understand the elements of an effective asset management system.

The organizational plan is the corporate plan, and is the critical, necessary, and vital first step to an asset management process. An organizational plan includes great detail about how the company intends to be successful. These 'details' are the objectives, and organizational objectives are relevant to what we are discussing. These objectives come from the strategic planning accomplished at the highest levels of the corporation.

The plan should be communicated in such a way that all parties are knowledgeable as to what the organization is all about and the impact it hopes to have in the space in which it operates.

Our company's organizational plans are:

The plan provides details for the objectives. These objectives, as well as others, help to define the trajectory of the organization and effectively point the direction the company must follow to ensure that all its assets (physical, human, financial, etc.) are working together (in harmony) to reach the goals of the organization.

The mission statement that guides our company should include a description of what our organization holds as its core values. Our core values and the objective evidence that we hold to these values include:

Before we can defeat our enemies we have to name them. The chief competitor in our space is:

The organizational objectives should be far-reaching and lofty, yet speak to the common man. ISO 55000 is intending for our organization to take the conversation of these core objectives down to the floor level.

Our company's organizational objectives are:

Similar to any policy, an asset management policy is a statement of intent. This intent should address the guiding principles that are important to the organization and be supportive of the organizational plan and the organizational objectives.

For our asset management policy, we will consider evidence that clearly indicates that:

- The policy was created by stakeholders
- The policy is followed
- The policy is reviewed from time to time to ensure it still supports the organizational objectives

The SAMP is the transformational document that converts organizational objectives to asset management objectives. A well-conceived

Strategic Asset Management Plan will also lay out the approach to be taken to comply with the overall asset management objectives.

The SAMP is an iterative process and is forever being evaluated and improved. An iterative process is a process that gets ever closer to the desired results the more iterations that are conducted, similar to the Deming Wheel.

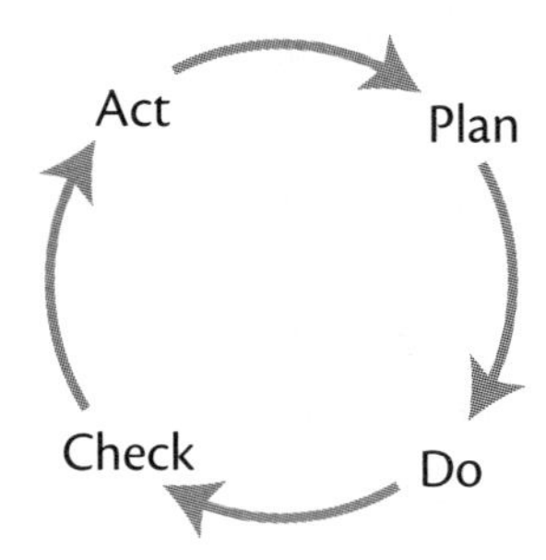

Our organization needs to show that there is constant reviewing and aligning of the SAMP and asset management objectives with the organizational objectives. These reviews become the objective evidence that the SAMP is a living product and is 'freshly' in tune with the direction of the company.

Our asset management objectives include:

The SAMP will be the guide for setting up the asset management objectives and describing how the asset management system will work to ensure the objectives are met.

Asset management plans are the tactically executed activities that an organization performs to match the asset's performance against the organizational objectives. This is actually the care and upkeep of the asset in a production or facilities setting. The activities we are involved in for asset management are the fundamentals of good maintenance and reliability listed here:

Our goal is to accomplish these fundamentals, consistently, at the highest level of execution.

ISO 55002 instructs organizations to develop these asset management plans for the purpose of defining the activities that will be implemented and the resources that will be used to meet the asset management objectives. ISO 55002 continues to suggest the plan include such detail as:

- Rationale for our activities around asset management
- Operational plans
- Maintenance plans
- Risk management plans
- Asset rebuild, replacement, and upgrade plans
- Financial and resource plans

The final stages of an effective asset management system include the measure of asset performance against the project value to meeting organizational objectives. A constant fine-tuning is required through a robust continuous improvement environment.

References

Carlin, Daniel. (2017, January 24). (Hardcore History-Show 59 (Blitz)). The Destroyer of Worlds [Audio podcast]. Time stamp 2:05:59.

Deming, W. E. (1986). *Out of the Crisis.* Cambridge, MA: Massachusetts Institute of Technology.

Gulati, R. (2nd Edition, 2013). *Maintenance and Reliability Best Practices.* New York, NY: Industrial Press, Inc.

Heinrichs, J. (Editorial Director) (January 2019). Our People. *Southwest The Magazine,* 22.

http://www.corning.com/worldwide/en/about-us/company-profile.html.

International Standard, *ISO 55000.* Corrected version 2014-03-15. Published in Switzerland.

International Standard, *ISO 55001.* First Edition 2014-01-15. Published in Switzerland.

International Standard, *ISO 55002.* First Edition 2014-01-15. Published in Switzerland.

Kiyosaki, R. T. (1998). *Rich Dad Poor Dad, What the Rich Teach Their Kids About Money—That the Poor and Middle Class Do Not!* New York, NY: Warner Business.

Liker, J. K. (2004). *The Toyota Way, 14 Management Principles From the World's Greatest Manufacturer.* New York, NY: McGraw-Hill.

Moore, R. (2004). *Making Common Sense Common Practice, Models for Manufacturing Excellence.* Linacre House, Jordan Hill, Oxford, UK: Elsevier.

Nanus, B. (1992). *Visionary Leadership*. San Francisco, CA: Jossey-Bass.

O'Connell, F. (2001). *Simply Brilliant*. London, U.K.: Pearson Education Limited.

Rogers, J. (2004). *Adventure Capitalist*. New York, NY: Random House Trade Paperbacks.

Ross, J. (2018). *The Reliability Excellence Workbook, From Ideas to Action*. South Norwalk, CT: Industrial Press.

Ross, J. (Jun/Jul 2009). *Establishing Proactive Maintenance*, Uptime: Fort Meyers, FL.

Ross, J. (2/2018). *Stock It? Don't Stock It?* MaintWorld Maintenance & Asset Management: Helsinki, Finland: Omnipress Oy.

Schroeder, A. (2009). *The Snowball, Warren Buffett and the Business of Life*. New York, NY: Bantam Books

Wikoff, D. J. (2015). *Leader's Guide To ISO 55001: Asset Management System Requirements*. Mt. Pleasant, SC: Institute at Patriots Point.

Wilson, A. (2002). *Asset Maintenance Management, A Guide to Developing Strategy & Improving Performance*. New York, NY: Industrial Press, Inc.

Index